A BIG BANG
IN A LITTLE ROOM

A BIG BANG
IN A LITTLE ROOM

THE QUEST TO CREATE NEW UNIVERSES

Zeeya Merali

BASIC BOOKS
New York

Published in the United States by Basic Books, an imprint of Perseus Books, LLC,
a subsidiary of Hachette Book Group, Inc.

Books published by Basic Books are available at special discounts for bulk purchases
in the United States by corporations, institutions, and other organizations. For more
information, please contact the Special Markets Department at Perseus Books, 2300
Chestnut Street, Suite 200, Philadelphia, PA 19103, or call (800) 810-4145, ext. 5000,
or e-mail special.markets@perseusbooks.com.

Designed by Amy Quinn

Library of Congress Cataloging-in-Publication Data
Names: Merali, Zeeya, author.
Title: A big bang in a little room : the quest to create new universes /
  Zeeya Merali.
Description: New York : Basic Books, [2017] | Includes bibliographical
  references and index.
Identifiers: LCCN 2016036546 (print) | LCCN 2016040738 (ebook) | ISBN
  9780465065912 (hardcover) | ISBN 9780465096619 (e-book)
Subjects: LCSH: Cosmology.
Classification: LCC QB981 .M465 2017 (print) | LCC QB981 (ebook) | DDC
  523.1—dc23
LC record available at https://lccn.loc.gov/2016036546

10 9 8 7 6 5 4 3 2 1

For my parents

Contents

1

God's Billboard:
The Cosmic Microwave Background

This is a book that is all about beginnings. It tells the story of the people uncovering the secrets of our universe's birth, scientists who have spent decades striving to understand the origins of space, time, our cosmos—and potentially many other parallel universes in an ever-inflating multiverse. Over the course of the following chapters, I will meet with them and others who argue that we could wield that knowledge to forge a new cosmos in the laboratory, with the help of some exotic physics, a few weird particles, a lot of energy, and a little bit of luck. And I will also be examining the ethics of whether or not we should perform such an act of creation, given the responsibility that comes with it. After all, once our baby was born, the theories say, it would evolve into a full-fledged universe, replete with new galaxies, new planets, and maybe even new life: a daughter civilization that would look upon us as gods.

So it's somewhat ironic that the book's beginning, this opening chapter of my journey, is starting off so badly.

Before committing to this (possibly quixotic) quest, I want to know if there's actually any point in entertaining the idea that we humans could become cosmic creators in our own right. Do any scientists believe we could truly make it happen?

In some sense, we cannot know for sure that we can make our own universe until we've done it. So I turn the question inside out: if we can, even in principle, knock together a baby universe, then it must stand to reason that our own cosmos might have been created by a more advanced civilization or superior intelligence—and we exist as proof of its success. If we could find evidence of that, it would be a huge discovery in its own right, reverberating through science and theology. And if our makers happened to leave any blueprints for how they performed this trick, so that we could repeat it, then all the better.

But does anybody, excluding science fiction authors, seriously think that the possibility that we could find signs of our alien makers (or, dare I say, a divine Maker, with a capital *M*) is plausible? And could we investigate this as a scientific, rather than metaphysical, hypothesis by searching for evidence? If so, then I think there is good reason to pursue this universe-building idea in depth.

The first person to ask, I decide, is Anthony Zee, a Chinese American physicist at the University of California, Santa Barbara. Zee not only has considered the possibility that the universe was made by an external intelligence but also has gone as far as calculating how to decipher any hidden communications scrawled across our cosmos from—as he and his co-author Stephen Hsu put it in their speculative paper around a decade ago—"some superior Being or Beings who got the universe going." Simply put, they figured out how to read the word of God, or that of our alien

Figure 1: The cosmic microwave background is the radiation that filled the universe just 380,000 years after the big bang. This map shows slight variations in the temperature of this radiation across the sky. In this grayscale image, lighter regions are slightly warmer; darker regions are slightly colder. Credit: ESA and Planck Collaboration

overlords—should such an entity or entities exist and have chosen to message us, that is—scribbled across the skies.

The place to hunt for this code, they claimed back in 2006, is in the radiation that pervades the sky—the cosmic microwave background, which is an echo of the big bang and would be the (hypothetical) creator's chalkboard, as it were. We're immersed in this radiation bath, we pass through it every day, but we're largely oblivious to it. It's worth taking the time in this chapter to come to grips with how this radiation came about because, aside from being a convenient place for a deity to leave us a memo, it also, less whimsically, serves as the strongest evidence of the big bang theory. And it turns out to be the best place to look for support for the model, which we shall meet later in the book, that provides the instruction manual for building our homemade universe.

I have traveled to Santa Barbara from London to meet with Zee, and I have a short list of questions to put to him about his paper with Hsu, tantalizingly titled "Message in the Sky." Chief among them: is this a joke? That was actually my first reaction to the paper when it was released as a preprint back in 2005. Back then, my job as a reporter at *New Scientist* magazine was to find newsworthy research, and largely involved spending hours trawling through arXiv, the physics preprint server where physicists post new papers for the attention of the academic community, often at the same time as submitting them to a reputable journal, where the papers will be reviewed by their peers before publication.

But arXiv is a murky place, where goofy papers stand shoulder to shoulder with respectable workaday reports and, very occasionally, groundbreaking results. Knowing that these papers have yet to pass peer review and be officially sanctioned, you have to exercise a bit of care before choosing whether to cover one of them. Many never make it to journals.

The title initially made me worry that the paper was either a deliberate spoof or bad science. I quickly dismissed both of those concerns when I saw Zee's name as an author, however. The physicist is a well-established senior researcher, with a long track record working out the kinds of things that would happen when you smash particles together in an accelerator, such as the Large Hadron Collider (LHC), in Geneva, Switzerland (just the kind of place where some physicists hope that a new universe could be born, in fact). Zee is not a crank.

Reading further, I could see that the science that he and Hsu outlined, which we'll get into in detail a bit later, was also perfectly sensible. Still, at the time, I decided not to write a news article about it, partly because of the worry that it would be too oddball a topic to cover. But it had stuck in my mind ever since. What I really wanted to know back then, and what I still want to know today as I rap my knuckles on his office door, is what was the authors' motivation for thinking about this question. In his core, does Zee honestly believe our universe was created by a God or gods, or aliens whose technological capabilities lie just beyond our own—and that we really could read off proof of this using experiments running today? Does that have bearing on whether he thinks we should carry out such a cosmos-making task ourselves? Or was the paper just an amusing example of an exercise all theoretical physicists revel in: asking fantastical what-if questions, just to see where they lead, without believing they could be literally true?

There are hints that Zee really might believe there's a guiding hand at work in the cosmos. He is an accomplished popular physics author in his own right, and in his books for a lay audience he has described physics as an almost spiritual search for beauty in nature—for, in his own words, the "ultimate design of the world." He's also characterized it as an effort to gain insight into the "mind of the creator," and he has pondered how stringently physical laws limit "God's" freedom.[1] But that's by no means definitive; physicists are prone to waxing lyrical about the elegant rules that mold reality and often use the term "God" as a metaphor for nature's underlying order.

Putting these questions to Zee during our meeting seems like a simple enough task. But moments after I step through his door, things start to go wrong. It rapidly becomes apparent that this will be a tough conversation—if I'm allowed to speak with him at all. Zee is a senior and highly respected researcher in his early seventies who, despite his age, still cuts a tall, severe, and formidable figure. He sports a full head of black hair, with only smatterings of gray, and his face is barely lined. I had emailed him ahead of my visit to tell him the topic of the book, and that I planned to interview scientists and also philosophers and theologians. This seems to have raised alarm bells. His tone is firm and humorless as he tells me, in an American accent that carries only a trace of his Chinese origins, "I want to make clear, I am not the new Messiah."

"I . . . no, no, I know," I stammer, and laugh nervously. This is a man who has worked with, and names as close friends, numerous Nobel-laureate physicists. He takes pains to tell me that he has done research and written seminal textbooks on perfectly sensible topics such as gravity and quantum fields, as well as having written a host of popular best-sellers. I realize that he already seems to think I am a fruit loop—some kind of religious zealot. He explains that the "alien communication paper" I contacted him about accounts for less than 1 percent of his usual activities. (Throughout our conversation, I later note, he never uses the term "God" when he discusses the "Message in the Sky" paper. Contemplating the notion that a god made our universe is several rungs on the wackiness ladder above the idea that it was made by aliens.) I suddenly feel very silly and self-conscious, as I am about to ask him a series of frivolous questions about reading messages from our universe's creator(s) and whether humans could one day become gods.

The trouble is that, unlike many of the people who will feature over the course of this book, I have never personally met Zee before or corresponded with him in any depth. Academic physics is a small world, and many physicists I come into contact with nowadays either knew me as an undergrad or doctoral student in cosmology or have met me since at physics conferences I have covered as a reporter, and have come to trust (and I hope maybe even like) me. Others know me as the editor of the Foundational Questions Institute (FQXi) website, run by an organization that doles out not-insignificant wads of cash for physics research that borders on the metaphysical, or they have read the writing I've done for various science magazines. But Zee is a mystery to me, as I am to him. He is the unknown variable (the x or, perhaps more fittingly, z, which is the sign for a complex unknown number) I have yet to solve for, and I can tell he feels the same way about me.

Zee steps out to make a cup of tea before the start of the interview, which I am fully aware is still in jeopardy if I make a misstep. While he's out, I quickly set up my voice recorder, to help me take notes. This, it turns out, was not a wise move. On his return, he takes one look at the recorder, and his posture stiffens. "I don't usually allow recordings," he tells me. "And I am now old enough and, I like to think, successful enough as a physicist and a writer that I can refuse interviews." While this might make him sound arrogant, he's not. I can see from the wariness in his eyes that he's just a brutally intelligent man who has been burned by his dealings with the media in the past. Indeed, he shows me an example of a piece in the Indian press that made him out to be the leader of a new physics-based cult.

I quietly rue the fact that I never learned shorthand, and explain that though I am taking longhand notes, the recording will help me keep his words straight. I promise I have no desire to wreck his reputation and I will not broadcast the recordings without his consent, nor will I present him as the Second Coming. He grudgingly allows the interview to begin—but when it does, it is he who is grilling me, rather than the other way round. In Zee, it seems, I have a hostile witness who not only does not want to answer my questions but is actively suspicious of me.

Zee's interrogation follows: "Where and what did you study?" Cambridge University, for my undergraduate and master's degrees (specializing in theoretical physics, first class) and then Brown, for my PhD in cosmology. No, I didn't focus on the cosmic microwave background (the divine message board) during my doctorate, but I did as part of my master's research project, I explain when prompted, hoping this scores me some points. I may be a crackpot, but at least I'm a highly educated one.

But it's Zee's next question that finally shatters the ice between us: "Whom did you do your PhD with?" Robert Brandenberger, I reply. The name sparks warm recognition, and Zee's entire demeanor changes. His strong, severely square face lights up in a broad grin and his shoulders relax. Brandenberger is a world-renowned cosmologist and sometime string theorist, a man who has quietly turned his back on establishment theories of the universe but has managed to do so with such refinement, politeness, and finesse that he has always retained the respect of his more mainstream colleagues. (Or, as Zee puts it, "Robert is not a wild man!")

More important, it turns out, is that my PhD supervisor was a postdoc at UCSB back in the 1980s, before Zee had a permanent position here, but while he was visiting for a year. At the end of Zee's stay, Brandenberger offered to drive Zee's belongings back to his permanent residence in Seattle, since Zee needed to head off on a road trip before he could go home. But when Zee eventually arrived back in Seattle, after meanderings to Princeton and New York, he was offered a permanent job back at UCSB. "So I never had to move that stuff [to Seattle]—that stuff didn't even have to be unpacked. That's the punch line of that story," he guffaws heartily.

A tap has been turned on, and the words, until now halting and careful, come spilling out. "I think I am going to like talking to you," he continues. "You and I are from the same world, roughly." My advisor's kind offer (which turned out to be unnecessary) has paid dividends for me some thirty years later. Suddenly I'm seated in front of an entirely new man—a natural raconteur with an anecdote for every occasion—whose wide grin has softened his face, dropping twenty years off his age.

Before we talk about the birth of the universe, Zee explains his own origins, which are possibly even more convoluted than those of our cosmos.

He was born in China in 1945, but his parents left for then British-owned Hong Kong after the communist takeover by Mao Zedong in 1949. Like me, he says, he thus began with an English education, which seemed odd for him at the time. His parents were "highly Westernized" and nonreligious, but his elementary school teacher was a Catholic and, after a visit to the Zees' house, inspired his mother to convert; his father soon followed. When Zee reached junior high age, the family migrated to Brazil, where his father, a businessman, donated money to build the first Catholic church in São Paulo for Chinese immigrants. The church still carries his late father's name, though Zee says his Catholic upbringing had little influence on his later work in physics. Indeed, he "let that go" after he came to the States to do his undergraduate education at Princeton.

It was on the forty-day boat ride to South America as a boy that Zee fell in love with physics. His parents had known little about what to expect on their arrival in Brazil, coming there as Chinese refugees. They could not even be sure that he would be able to attend school. So his mother had bought him a pile of high school textbooks, which he devoured on the boat, with physics and math hooking him the most. Their fears were unwarranted, as he was able to attend an American international school in São Paulo. As a consequence, he says, he knows little Portuguese to this day, because the school taught only English and French. ("What do you want to learn Portuguese for?" the principal had told one of Zee's fellow students, who had asked about studying the language. "To address the servants?" Zee winces as he recounts the parochialism.)

By good fortune, Zee's family home was near the American consulate, where he read popular physics books, his favorite being *One, Two, Three . . . Infinity*, by Russian physicist George Gamow. Little did the very young Zee know it back when his parents bundled him out of China, but 1949, the year that saw the communist revolution, also saw the universe being overhauled, in part due to Gamow. That was when the term "big bang" was coined to describe how our cosmos began, and Gamow, one of the major architects of the model, proposed that this explosion of space and time would have left behind a bath of radiation, still detectable today, which he dubbed the "afterglow of creation." This is the cosmic microwave background, which would later fascinate Zee.

It's worth taking a bit of time here to establish how the physics of the preceding century had led physicists to the big bang model and the prediction of the cosmic microwave background. The theory behind building a universe in the laboratory is an extension of the ideas that had been set out by 1949, and that extension—as yet to be definitively proven—takes massive credence from precise satellite measurements of the background radiation.

I'll follow Zee's lead here in describing that history, as outlined in his first popular physics book, *Fearful Symmetry*. (The name of the book derives from William Blake, and the reason for this choice, and how Zee's own penchant for a poetic turn of phrase in that volume has since seen him mistaken for a physics guru, will become apparent along the way.) In that book, Zee traces the roots of the big bang to the early nineteenth century and ties them, perhaps surprisingly, to the invention of the battery by Count Alessandro Volta.

The battery, Zee notes, gave physicists a canned source of electricity to play with—and play they did. By 1819, Hans Christian Oersted had discovered that if you pass an electric current through a wire, it causes a compass needle nearby to twitch; electricity could create a magnetic effect. On the flip side, English physicist Michael Faraday later found that a magnet moved through a coiled wire generates an electric current. Electricity and magnetism, it was quickly realized, are two sides of the same coin, and the discipline of electromagnetism, combining the two, was born. Faraday expressed these phenomena in terms of the existence of electric and magnetic fields, emanating from either an electric charge or a magnet, respectively, that could act on objects near the source of the field without directly touching them. Adding the final piece to the puzzle, Scottish physicist James Clerk Maxwell came up with a set of mathematical equations to relate the two kinds of fields.

The key part of Maxwell's equations, for our story, is that they showed that a changing electric field in space will induce a changing magnetic field in its path, which in turn generates a changing electric field, and so on, creating a propagating electromagnetic wave. That wave, he recognized, was light. The speed of that wave is a constant that could be calculated from his equations and closely matched the measured speed of light that physicists had found in the lab. With one stroke, the fields of electromagnetism and optics were unified.

This is where Albert Einstein comes in—as he is wont to do in physics books. It is oft told that when just a sixteen-year-old boy, Einstein pondered what would happen if you chased a sunbeam and caught up with it—something like how the young Clark Kent in Smallville could, on foot, keep pace with the bus taking his friends to school in the morning.[2] What, Einstein asked, would you see? Objects are visible to us only because light bounces off them into our eyes. But if you were running alongside a light beam, what would it look like? Would the light beam freeze in front of your eyes?

Einstein had studied Maxwell's work and so he knew that the speed of light was a fundamental property of electromagnetic fields. When Maxwell had come up with his equations it seemed sensible to think that light must be transmitted through some kind of medium. That, after all, was how other familiar waves travel: sound, for instance, is carried by the compression of air, and water waves are undulations on the surface of a liquid. The equivalent hypothesized carrier of light was dubbed the "luminiferous ether" (literally "light-bearing medium"). But no experiment or astronomical observation was able to find evidence of this substance, which presumably should have been influenced by the motion of Earth around the Sun and the movement of the solar system through the galaxy. If this were the case, you might expect that the time it took starlight to travel to Earth would vary depending on whether Earth was moving toward or away from the star while on its orbit, and how, as a result, it dragged the ether, but this effect was never seen.

This, along with other lab experiments, led Einstein to reason that there was no ether and that the speed of light moving through the vacuum of space must be constant. In other words, there was no way to catch up to it, no matter how fast you run; it always whips away from you at nearly 300,000 kilometers per second. He enshrined this principle in his special theory of relativity, published in 1905. But if light's speed is set in stone, something else has to give, to account for why you could never gain on light. That something else, Einstein realized, was both time and space.

To understand why, imagine that Lois Lane tries to carry out a series of experiments, with Clark Kent as her lab assistant. She instructs Clark to run at 200 km/hr while she serves a tennis ball at 300 km/hr in the same direction that Clark is running. At what speed, she asks, does he measure the ball traveling ahead of him, as he sprints? Clark replies that the ball appears to be moving with a relative speed of 100 km/hr (300 km/hr − 200 km/hr). If he speeds up to 300 km/hr, then the ball is moving directly alongside him and appears stationary. Clark has successfully caught up with the ball.

But Lois knows that, according to Einstein, things become tougher when you try to chase after light, rather than a ball, because the speed at which light moves away from you remains a constant, no matter how fast you are. Challenging Clark further, Lois now shines a flashlight, knowing that the light will move away from her at 300,000 km/s, and tells Clark to run at 200,000 km/s and measure the speed of light that he sees. But now, he says, rather than seeing the light slow to 100,000 km/s (300,000 km/s – 200,000 km/s), he sees it still running away from him at 300,000 km/s.

What's going on? Einstein's insight was that the only way for both Lois and Clark to measure the same speed of light was if they each make different measurements of how much time has passed or how much distance is being covered. This difference is not because one has bad equipment or the other cannot count properly. It is because time passes at different rates and space is distorted for people in motion relative to each other. From Lois's point of view, after one second has passed, the light has moved 300,000 km and Clark has only moved 200,000 km from her, so she would argue that from her perspective, the light is 100,000 km ahead of Clark (300,000 km – 200,000 km).

Until Einstein suggested otherwise, people had rather sensibly assumed that clocks (at least, ones that are working properly) run at the same rate—1 second per second—for Lois, Clark, and everyone else anywhere in the universe. But what if time passes more slowly for Clark while he is in motion—that time passes at, say, one-third of the rate measured on Lois's stopwatch? Then to him, the light has traveled that same 100,000 km ahead of him that Lois noted, but in just one-third of a second according to Clark's watch, rather than taking one whole second, as measured by Lois. So Clark would reason that light's speed is still 300,000 km/s.

Another possibility is that a discrepancy between Lois's and Clark's measurements of the distance between Clark and the head of the light beam is what causes the mismatch in their conclusions. In fact, it is a combination of both these factors: clocks tick more slowly for moving observers (an effect known as time dilation) and space is compressed along their line of sight (a phenomenon called length contraction). These predicted effects seem bizarre to us because they only become noticeable at speeds well beyond those of our everyday experience, but they have been confirmed experimentally. Time dilation has been tested by comparing the time that elapses on atomic clocks placed on supersonic jets with synchronized clocks left on the ground. Evidence for length contraction has also been spotted by physicists studying the paths of exotic particles called muons that travel through Earth's atmosphere at close to the speed of light. Sitting in labs on Earth, these muons decay far too rapidly to travel a distance as great as the thickness of the atmosphere. But when hurtling toward the atmosphere at high speed, the distance they have to cover contracts, while the passage of time slows, allowing them to make the trip through.

Einstein's theory forces us to accept that time is another dimension, alongside our three spatial dimensions. Just as two people standing at different positions in space in Zee's office might disagree about whether Zee's teacup is placed squarely in front of him on the table or sits slightly to his left or to his right, two people can also disagree on whether two events—Zee sipping his tea and my placing down my pen—occurred simultaneously or not; if not, then those observers can argue about which happened first. Neither is wrong; their perception of the ordering of events simply varies depending on their motion through spacetime.

Special relativity is dear to Zee's heart. He cites it as a perfect example of symmetry and beauty in physics. Symmetry is, in some sense, beautiful to everyone. If you're born with a symmetric face, scientists say, you've lucked out in the gene pool, as this is supposed to make you more attractive. (As someone adorned with a slightly wonky nose, I am somewhat bitter about this claim. I guess you could say it has put my metaphorical nose out of joint, to match my actual one.)

Physicists also see beauty in the symmetry of nature—it is often a guiding principle in their formulation of theories. They mean something slightly more subtle, though, when talking about the rules that govern reality. You can say a square has symmetry because if you turn it by 90 degrees or flip it over, it looks the same. Similarly, physicists seek out symmetries in the laws that make up the world, such that even if you change the situations in some regular way or view it from a different angle, certain aspects of physics still apply in just the same way.

In special relativity, the symmetry lies in the fact that despite their relative motion, all observers measure the speed of light to be constant. This is not quite as arbitrary as it sounds. (Why is the speed of light so important? Why isn't the speed of sound constant? Or the speed of badgers?) Remember that Maxwell's equations had shown that the speed of an electromagnetic wave through a vacuum depends on the properties of the electric and magnetic fields that induce each other, in a self-propagating chain. These properties do not change for the two observers, even when they are moving relative to each other, and thus light's speed should be fixed. (By contrast, sound travels by physically compressing the medium that it passes through, and so it is tied to the properties of that material. The speed of badgers, meanwhile . . . well, I'm not a biologist, but I guess it depends on how hyped the badger is feeling before the race.)

It's the balance and order in the world that lie at the heart of Zee's first book, in which he enumerates our most fundamental physical laws that obey symmetry principles. This common factor between laws enables physicists—more so than those dabbling in the other sciences—to create models of nature that do not always have to be based on experimental evidence, in the first instance, but can be conjectured by closing your eyes and just imagining purely "what if?" If a certain symmetry is preserved by nature, what laws would arise? Of course, for these models to pass from fantasy to reality, you still have to go out and make measurements of the real world to check they are correct, but it is a topsy-turvy way of doing science. Now you can see why Zee chose to title a book about the awe-inspiring nature of the world *Fearful Symmetry*, borrowing from William Blake. "You wouldn't have a book like that about chemistry," remarks Zee with a smile.

But while many physicists identify symmetry and elegance as essential keystones in building physics models, few would employ the loaded terms that Zee uses freely in his book. "I had some quasi-spiritual-philosophical statements in there," Zee admits, pulling the first edition from the shelf in his office and flicking through it, pointing some of these out to me: he talks about understanding the "mind of God," the "ultimate designer of the world," and the "creator." It's perhaps not so surprising, then, that the Indian journalist he'd mentioned earlier mistook him for a physics guru, expounding a spiritual path to science, but Zee cringes at the memory.

Even before publication, Zee tells me, he was worried he would be denounced as a kook. He'd signed a contract to write the book just before receiving the permanent job offer at UCSB, after being encouraged to write for the public by one of his mentors, Nobel-winning physicist Steven Weinberg, himself a prolific popularizer of physics, who admired Zee as a wordsmith. But when it came to publication, Zee was terrified that the book would ruin his reputation and that people would feel he could no longer perform research. That fear did not pan out, yet it seems to still hang over him during our interview, and he remains guarded as I try to coax him to speak about whether he actually holds any spiritual beliefs.

I point out that Zee did in fact use that language in his book, despite knowing the kind of attention it might draw. It seems odd that somebody so smart, so literate, and so aware of the effect his words might have still chose to use those terms. Did he believe what he was writing? Or was he simply being a provocative mischief maker? "You can just say it is poetic talk," he says.

The book opens by quoting Blake's verse: "Tyger! Tyger! burning bright / In the forests of the night, / What immortal hand or eye / Could frame thy fearful symmetry?" "The nature of the tiger is defined by the nature of God. But then it is God who is defined by the nature of the tiger," Zee explains. "And so I like to think of the ultimate designer defined by . . . " But then he trails off, leaving me to fill in the gap: the ultimate designer of reality is defined by the structure and symmetry in the world.

Still, he's hedging. You may not wish to answer, I say, but do you literally believe there is a God? Or are you equating "God" in a metaphorical sense with the beauty and symmetry of nature? Zee pauses for a long time, and then answers carefully that as he gets older, he finds it increasingly "depressing" when physicists make statements that we will soon know everything there is to know about the universe. With a sigh, he quotes a passage that Weinberg—a strident atheist—famously included in his own popular book, *The First Three Minutes*: "The more the universe seems comprehensible, the more it seems pointless."[3] "I find Weinberg's statement very depressing," Zee says.

But Zee clarifies that this does not mean he is a theist. "Given our present understanding of the universe, I think it does seem difficult to believe there is a personal God who is interested in *you* personally, given the vastness of the universe," he says. He adds that he is almost certain, given the huge size of the universe, that aliens exist too, giving God even more intelligences to watch over. "I think any informed and scientifically trained person would have to say that. And this personal God—I mean, it is possible," he says with a laugh, "but He would have to be interfering in everyone's life. So I don't know. I have to think about this."

That's fair. But for the record, I do believe in a personal God. Zee's reply does not dissuade me from this belief, simply because I think once you've gone partway to accepting that a supernatural intelligence can create space and time, you may as well say that, lying as He does outside time, it is probably fairly easy for Him to multitask. (Indeed, in Chapter 2, we'll come to a physicist who controversially claims that quantum experiments of the microscopic world already support this view.)

Regardless of its origin, this intrinsic orderliness, beauty, and symmetry in the world also drove Einstein to develop his general theory of relativity. By 1905, he'd published his treatise on the special theory of relativity, asserting the constancy of light's speed and the malleability of time and space, which together made four dimensions. In 1915, he followed up with his general theory of relativity, which brought in gravity (and spectacularly belied the rule of thumb that sequels are never as good as the originals).

The symmetry guiding Einstein in the context of general relativity is a little bit harder to see than in the case of the special theory, where all observers must measure the speed of light to be the same, no matter their relative motion. To be more specific, the observers in that theory are moving at a constant speed relative to one another and are not changing direction—in other words, they are not accelerating—and there's no real need to take gravity into account to explain their motion. That's the "special" part of the theory because it deals with a quite constrained condition. But in general, things do speed up or slow down relative to other objects, they often change direction, and they are shackled by gravity—especially on cosmic scales, where gravity holds planets in the grip of the stars that they orbit. How, Einstein wondered, do you take acceleration and gravity into account? The symmetry that he picked out, in a thought experiment while at work at a (clearly deeply boring) patent office, was that acceleration and gravity are equivalent.

The classic example for seeing this is to imagine yourself in an elevator on Earth that is not moving up or down. If you take out an apple and drop it, you'll see it fall to your feet—pulled down by Earth's gravity, and moving faster and faster as it falls, accelerating at a rate of 9.8 m/sec/sec. (Don't forget that centuries earlier Galileo had shown that anything dropped off the tower of Pisa would fall at the same rate, if you discount air resistance: apples, oranges, bowling balls, or graduate students.)[4]

Now, if that elevator was transported into deep space, away from any huge planets, and you were floating like an astronaut in free fall, the situation would look very different. There's no way to mistake these two situations for each other—unless, that is, you imagine the elevator doing what it was designed to do and accelerating upward, let's say at 9.8 m/sec/sec. If that was the case, you would not be experiencing free fall. Instead, the floor would be pushing up at your feet and you would feel as though you were being pulled down, just as you are on Earth. Take out your apple and drop it again, and crash, down it goes. The two situations are symmetric: gravity and acceleration are equivalent.

Einstein already knew that time and space are flexible and that motion distorts them. So it seems likely that acceleration would be bound to jerk around with both of them too. Since he could now see that acceleration and gravity are equivalent—that is, you can describe the effects of gravity by looking at a situation far away from gravity in which your system is accelerating—then gravity must pervert space and time as well. To describe this warping effect of gravity, Einstein knitted the four dimensions together into one spacetime fabric pervading the universe.

Gravity, Einstein argued, manifests because heavy objects warp spacetime, causing it to bend and curve around them. The iconic image is that of a rubber sheet with a heavy bowling ball added, stretching the sheet downward. Throw an orange onto that sheet and you will see that it veers toward the heavier object. It looks as though an invisible force, gravity, operates between the two, pulling the lighter sphere toward the heavier one. In fact, the orange is simply following the contours of the deformed sheet.

Light should also trace the shortest route through this rolling spacetime, causing its path to bend as it passes massive objects, following spacetime's contours. Einstein's equations that accompanied his theory predicted just how much distant starlight should swerve on its way around the Sun to Earth—and this prediction was confirmed in 1919, by English physicist Arthur Eddington's team, who measured the angle of that deviation during a total solar eclipse from the island of Principe, off the coast of Africa. Einstein's theory was vindicated, along with its startling assertion, succinctly put years later by Princeton physicist John Archibald Wheeler, who had worked on the Manhattan Project and then general relativity in the 1950s, as "Spacetime tells matter how to move; matter tells spacetime how to curve."[5]

Einstein's equations also predicted something else startling, something not even he was comfortable with: the universe must be dynamic. He, like others at the time, had thought of the universe as spherical and static, neither expanding nor contracting. His math told a different tale, however, which was that the universe was poised in a delicate balance, like a pencil standing on its point, between the amount of matter it contained and the amount of radiation, or light, held within. Nudge it even slightly, and the universe would tip, either expanding outward or shrinking inward. A young Russian physicist and meteorologist, Alexander Friedmann, took this idea and ran with it, creating a set of models for an expanding universe, whose fate depended on the amount of mass it contained: too much and it would eventually turn back in on itself, too little and it would expand forever. These two extremities were demarcated by a certain critical mass.

Still wedded to the static picture, and to maintain the status quo, Einstein threw in a stabilizer, a fudge factor that would counteract any tendency for the cosmos to collapse under the pull of the matter it contained by pushing space outward. This stabilizer, dubbed the cosmological constant, saved the static picture; the only trouble was, it was wrong.

By the 1930s, American astronomers Vesto Melvin Slipher and Edwin Hubble had measured the motion of distant galaxies and discovered that they were moving away from Earth. On top of that, the farther away they were, the faster they were moving. In other words, the universe was shown to be expanding. On the basis of this, Einstein denounced his introduction of the cosmological constant into his equations as the biggest blunder of his life. (Spoiler alert: It was not such a huge mistake after all. By the late 1990s, astronomers had reason to believe that such a constant does exist, and that it serves as dark energy, pushing space outward at an ever faster rate. But that's a story for Chapter 7, when the search for an explanation for the cosmological constant drives the quest to work out how to build a baby universe in the lab.)

The big bang model started to take shape as Roman Catholic priest and physicist Georges Lemaître imagined what it would be like to turn the clock back on this expanding universe, watching it shrink down further and further into a primeval atom, a dense ball of matter from which our cosmos was born. Friedmann's student George Gamow—who would become Zee's childhood inspiration—and colleagues put the pieces together to show that if the universe was born in a hot state, dominated by radiation, matter could form spontaneously. This would populate the infant cosmos with some basic chemical elements, and as the universe grew and cooled, this matter could eventually clump together to create stars, galaxies, planets, and, at some point down the line, the ingredients for people.

It was actually Gamow's student Ralph Alpher (who had worked with him on showing how matter could be created in the early universe) and Robert Herman who, in 1948, predicted the existence of the cosmic microwave background, not so much as the smoking gun of the big bang model but as the gunpowder residue left behind by the explosion. So what exactly is that background?

In some sense, the cosmic microwave background is a selfie that the universe took of itself in its childhood. Just after the zero hour of its birthing moment, both matter and light were created. Conditions in this cosmic furnace were so hot and dense—with matter tightly crammed in an unimaginably small cosmos—that light couldn't jostle past the matter and stayed imprisoned in this fiery ball. The same effect is seen today in the Sun. The sunlight that actually reaches Earth is released from the Sun's relatively cool outer surface, where the density of hydrogen and helium gas is low. By contrast, we cannot see light from the star's supremely hot inner core because the extreme density of the gas there traps the light inside.

In the first fractions of a second after the big bang, the temperature in the universe was a ferocious 10^{32} degrees Celsius—where 10^{32} is equal to 1 followed by 32 zeroes.[6] That's billions of times more blistering than the energy released by an exploding star, also known as a supernova. But as the universe expanded, it gradually cooled, and some 380,000 years into its existence the temperature was finally low enough that matter would have diluted sufficiently for light to escape. At that moment, the proverbial switch would have been flipped in the universe, bathing the cosmos in light.

That light burst was so bright that it is still detectable today—although we're not being blinded by this all-consuming flash. That's because as the universe expands, it stretches out light waves themselves, turning them from visible light into radio waves and then microwaves. Today, the big bang's echo is heard in microwaves that create a radiation bath that pervades space and has a temperature of around 2.7 Kelvin—that is, a chilly 2.7 degrees Celsius above absolute zero (the coldest theoretical temperature, where even atoms would stop moving and freeze in place), which is -454.5 degrees Fahrenheit. As much as 99 percent of the universe's light is trapped in this microwave glow, with a mere 1 percent making up the light of stars.

In 1949, this exploding-universe model was christened the big bang theory, ironically by the atheist physicist Fred Hoyle as a pejorative. It did not help dampen enthusiasm for the model, however, which ended up being blessed by Pope Pius XII in 1951 as a confirmation of the church's teaching that the universe had a beginning and was created.

Gamow, Alpher, and Herman called on radio astronomers to look for evidence of this mother of all lightning flashes, but they were told time and time again that it could not be seen. Those astronomers were wrong, however, as all that was needed to detect the glow from the beginnings of the universe was an ingenious piece of machinery: an old-fashioned analog television set. Before the digital age, if you tuned your TV to somewhere between channels, it would show the faint hum of white noise or snow. One percent of that snow was actually made up of radiation left over by the big bang.

The cosmic microwave background was first detected not through TV, however, but by accident, in another manner, in 1964. This Nobel Prize–winning discovery was almost dismissed as being caused by pigeons, when two engineers working at Bell Laboratories in New Jersey, Arno Penzias and Robert W. Wilson, spotted its effects while out on a routine mission. They had built one of the most sensitive microwave receivers to date and wanted to scan the skies for background radio signals that might interfere with radio broadcasts. Regardless of the direction in which they pointed their receiver, they picked up a faint microwave hiss. At first the pair assumed that there was a problem with the receiver. They noted that the instrument was filled with "white dielectric" material—bird droppings. But after clearing off the unpleasant guano and shooing away the offending pigeons, they found that the hiss remained. It was only when the duo read about cosmologists' predictions of just such a signal that they realized the enormous import of their findings.

A few years later, physicists Stephen Hawking and George Ellis at Cambridge and mathematician Roger Penrose at Oxford University showed that not just space but time must have been created at the big bang.[7] It was $t = 0$.

So by the mid-1960s, the big bang model—not quite in the form in which we understand it today, which we'll come to in later chapters, but with its essential features—had been confirmed. The universe began in a hot, dense flash, which brought forth both time and space, and later gave rise to matter. And the cosmic microwave background (potentially our creator's billboard, if Zee and Hsu's paper is right) had been found, all around us. There will be refinements to come, but for now, let's leave it at that.

Meanwhile, Zee, now a young adult, had been drawn to Princeton as an undergraduate, attracted by the thought of working with Wheeler, the charismatic physicist who so wittily encapsulated the meaning of general relativity. At Princeton, Zee became one of about ten Asian students, with maybe another ten minority students ("mostly African princes") in his class of roughly eight hundred—all male, and overwhelmingly white.

Zee tells me he did not suffer racism because at the time, with so few other minority students around, he was perceived more as a "pleasant curiosity" than as a threat. In many ways, he says, he found it easier to assimilate into US culture than he might today. He notes that now, as he arrives each day at the UCSB parking lot by the engineering department, he sees that most students are Asian, and in the canteens you find cliques where Chinese students gather to speak in Chinese, Indians at another table, Koreans in a different corner. "Nowadays you can practically go through an American university education [and] spend most of the time speaking your own language," he says. "By necessity, my generation merged into American society much more readily."

It was at Princeton that Zee began to move away from Catholicism, although back then he still had to attend chapel every Sunday because religious observance was mandated by the university. He recalls how the students had cards that had to be stamped, proving attendance. "You cannot imagine nowadays such a thing in American universities," he laughs. "But then the change in American society just occurred like this," he says, clicking his fingers. The catalyst? The Vietnam War. By then, Zee was in graduate school at Harvard, studying for a doctorate with Sidney Coleman. Alongside his classmates he participated in the antiwar riots in Harvard Square, so his most vivid memories of grad school are predominantly of police on horseback swinging clubs and releasing tear gas rather than of lessons about gravity or particle physics. "So American society was turned upside down," says Zee. "How are you going to make these kids go to church every Sunday?"

Zee excuses himself for a short break—the tea he has been sipping throughout the interview has taken effect. While he is out of the room, I think over his paper with Hsu, about a message encoded in the microwave sky. Since the first detection of the microwave background in 1965, there has been a series of ever more precise experiments sent up to map the sky more fully. As I mentioned, the rough temperature of the radiation is 2.7 K. But not all the sky is at that exact temperature; there are tiny fluctuations (just one part in a million) above and below that average temperature, which is what many of the experiments that measure the cosmic microwave background have been recording.[8] (I will explain the hypothesized origin of these wrinkles, thanks to the slightly jittery way that the early universe grew in size, in more detail in Chapter 4.) It is from these tiny wrinkles that astronomers read off features of the early universe and try to extrapolate its earliest moments.

These temperature ripples, in turn, correspond to slight wrinkles in the otherwise smooth density of matter in the infant universe's first moments, which mark out where matter began to clump together, giving rise to the galaxies, stars, and planets. The layout of galaxies and galaxy clusters around us today can be traced back to those ripples in the radiation—and by matching the patterns to predictions of how the radiation should look if the universe had a variety of different ages, astronomers have come up with their best estimate for the age of the universe: 13.8 billion years.

It is also within these ripples that Hsu and Zee suggested that a "creator," if one exists, could have encoded a message to be observed by humanity. Others have suggested that a divine message could have been recorded within the human genome or—as did Carl Sagan in the novel *Contact*—in the endless digits of pi (although, sad to say, no such code has yet been found). But the cosmic microwave background, Hsu and Zee argue in their paper, is the only physical "billboard" that lights up the whole universe and could be seen by any civilization, no matter where it resides in the universe. It's therefore the ideal place for a creator to leave a structured message, hidden among the otherwise seemingly random wrinkles, if you know where and how to look for it.

An important point, they also reasoned, is that such a code could only have been left by someone outside the universe and present at its birth. It can never be mistaken for one written by even the most advanced alien civilizations born elsewhere within this cosmos.

To understand why, you simply have to appreciate the vastness of this universe. We are limited to seeing the visible universe, the edges of which are set by the amount of time it takes light to reach us. Today the universe is around 13.8 billion years old, which means (at least naively) that the farthest objects we could be able to see today, out to the northern horizon, are 13.8 billion light-years away.[9] There may be other exciting things that lie beyond that, but light from those objects has not had time to reach us yet. Similarly, if you look out to the southern horizon, you could not see past 13.8 billion light-years. That means that these outer edges are separated from each other by 27.6 billion light-years, which makes them too far apart for any alien civilization that evolved *within* the universe to have scribbled the same message across the entirety of the sky—there just has not been enough time since the universe began 13.8 billion years ago for those aliens to zoom around spraying graffiti across the entire panorama.

The only time it would have been possible to engrave the entire sky with a single message was at the birth of the cosmos, when an external creator could have imprinted a message into the tiny seed from which the universe was born. Think of it like drawing a smiley face in marker on a balloon straight out of the package. Blow up the balloon and the picture stretches with the rubber. In the same way, as the cosmos rapidly inflated, its creator's message would shine out across the whole sky.

Hsu and Zee argue that the code could be simple binary, a series of 0s and 1s, represented by either a slight positive or negative deviation from the average temperature, respectively. You just need to scan the sky in enough detail to see if there is a meaningful message hidden among the otherwise seemingly random fluctuations in temperature. They reason that it would be possible to encode a message of about 100,000 bits—less than the RAM of a 1980s home computer such as the Commodore 64. That's not enough space to store the contents of your smartphone, but it could be enough to give an incontrovertible indication that the communication was deliberate. For instance, it could describe the known laws of physics that govern the maker's universe—something a daughter civilization advanced enough to have detected and measured the cosmic microwave background would know and recognize.

Zee steps back into the room, and I am about to ask him what gave him the idea for the paper when I realize that the tap that left him talking freely earlier—the one that changed him from stern and reserved to open and friendly—has somehow been turned off while he was out. "I think I need to stop now," he says as he walks in. It's suddenly time for me to go.

"I haven't asked you about your paper yet," I protest, surprised by this mood swing. Maybe he is just tired; we have been chatting for over an hour. But he seems uncomfortable and guarded again. Perhaps while he was out he figured that he was being drawn to talk a little too much about the boundary where physics crosses into metaphysics, and maybe even into religion.

"There's not much to say about the paper," he says abruptly.

This is an anticlimax. I start to pack up, but I hope to get some answers about the work while I do so. I ask him how he thought of this as a topic to work on. In an afterword added to a later edition of *Fearful Symmetry*, published in 1999, Zee had mused that our universe could be a "school-assigned science experiment carried out by a high-school student in a meta-universe."[10] Even so, it seems like a frivolous subject for investigation in an academic paper by such a serious-minded physicist, especially since it is so far removed from his usual research. He's still uncomfortable, and mumbles that though this is a notion he had been mulling over for more than thirty years, the idea to tackle the question rigorously probably arose during a chat with Hsu while the two were taking a walk.

I'm standing up now, ready to be ushered out, and I ask the main question I traveled to his office for: He has written about finding evidence that our universe was created by an alien civilization or by God. How much of that was tongue-in-cheek? Does he think it's a plausible idea? "I think it is a serious possibility," he replies. "And if the data is out there, why not, right?" In fact, he adds, a German colleague who specializes in data mining, Michael Hippke, is already thinking about carrying out this search.

Okay, I think I have my answer, then. Serious scientists are investigating whether our universe could have been created by superior beings, so it is reasonable to think that—one day—we humans could become advanced enough to perform the same feat of creation. This universe-building project is worth pursuing.

I hand Zee back his copy of *Fearful Symmetry* and, remembering his poetic turns of phrase about the nature of symmetry, add in passing that I hoped to ask him more about whether he thinks of physics as a spiritual search, in terms of seeking out beauty. "Well, that's a very good question because I am very much interested in that too," he says, adding that he has been toying with the idea of writing a book about it himself, but—as with the publication of *Fearful Symmetry* thirty years ago—he is worried about going out on a limb.

Zee is opening up again, but only cautiously. He volunteers that as he has become older, he has started meditating. He brings me back to my earlier question about his religious beliefs, which he says he did not quite fully answer on the first pass. In truth, although he did largely abandon his interest in Catholicism at Princeton, he has started to go to church again—a Methodist church, chosen because it is close to his house rather than for denominational reasons. Zee married for the second time ten years ago and now has a four-year-old son (his third child) with his new wife, and for him at this point, churchgoing is about giving his son a sense of community and religious instruction; during the services, Zee tends to meditate and "do his own thing." (Since our first meeting, when this conversation took place, I have spoken to Zee again, and he informs me that he no longer attends church.)

"I just feel very uncomfortable with the sort of totally rationalistic point of view that people like Weinberg express," he says, harking back to his old mentor's stance that the universe is ultimately meaningless. "It seems to me that there has to be a lot more than that," Zee continues. "But to express this view in academic circles is almost the kiss of death."

I see why Zee was so wary to speak about this now. This general attitude in the physics community has kept him quiet for years, and even today it seems to be preventing him from speaking freely about his "Message in the Sky" paper. He adds, without my prompting, that while they were preparing that paper, he asked Hsu to remove much of what he had written about the implications of such a finding, if it should ever be observed. Zee says that he regrets that now, but he does not elaborate on what exactly he censored. I make a note to call up Hsu and ask him if he remembers.

As we step into the corridor, a somber Zee recounts an incident that has stayed with him since he was a young professor in his thirties. A world-renowned senior physicist he had not previously mentioned—he tells me the name but asks me not to make that public—explained to him his view that there was no such thing as "mind," that your mind is just your brain, and your brain is nothing more than electrons interacting electromagnetically, and so humans have no free will and everything is determined. "I just said, 'I cannot believe that that is all there is!'" says Zee. His companion responded with a growl, "Tony, you're a physicist, what are you talking about? What else is there?" "When you're a young guy . . . with somebody like that, you just felt there's no way to argue back," Zee says. "Even decades later I feel very uncomfortable about that conversation."

On my way out, I ponder that I have been relatively lucky with my own academic career in that my religious views were never openly ridiculed or challenged in that way by my colleagues. Yet there is a bias in the physics community against even a slight wavering toward religious belief—even one as watered down as Zee's. So although Zee tells me he suffered little racism coming to the United States as a Chinese immigrant, that does not mean that he was free from the effects of prejudice.

Back in London, I put in a quick call to Steve Hsu, who is now vice president of research at Michigan State University and hence a man with a busy calendar. Nonetheless, he seems keen to chat, replying to my interview request within minutes. When we do speak, he tells me jovially that he's talking while cycling on his exercise bike; nowadays his admin duties fill so much of his schedule he has to "steal time" for research activities.

Hsu is happy to fill in some of the blanks from my conversation with Zee. The idea that a message from our universe's maker (or Maker) could be daubed across the heavens first hit him when he learned about the cosmic microwave background as a grad student at Berkeley and realized it would provide a unique way for someone present at the moment of creation to communicate to all its future inhabitants. "It's all based on a theoretical observation that everybody sees the same sky," he says. Hsu ran the idea past a few people, but he admits they were not particularly interested. "If you don't have a certain science fiction bent, then you'll just think it's nutty and not worth considering."

Around the same time, he came across some papers by cosmologists about how to build your own baby cosmos—laying down the first crumbs of the trail that would eventually lead me to write this book. Again, Hsu's interest was piqued. "It's like, these guys have actually shown you how to make a universe in the laboratory, and once you think that's possible, then you might ask, 'Oh, gee, can you send a message to the people that are inside?'" Strangely, he notes, this idea of starting up a universe was not widely appreciated by most physicists back then: "I went through it carefully at the time, but most theoretical physicists were just not even aware it was possible."

This strikes me as an important point. When I first came across the universe-building project in 2006, much later than Hsu, I too was surprised not to have heard of it before. Given its implications, why had it not garnered the attention that I—and Hsu, it seems—felt it so obviously deserved? It's something I will have to dig into as I uncover the history of the project, but I suspect it is in large part because the project's authors were as wary as Zee of being thought crazy, and played down the implications of their research to avoid unwanted attention.

It was only years later, once Hsu became a young professor at the University of Oregon, in Eugene, that he mentioned the idea to Zee, a long-term collaborator on other, more sensible topics in particle physics and gravitation. Zee, who for decades had whimsically contemplated the notion that our universe was created by an alien intelligence, saw its potential, and they wrote it up.

Hsu vividly recalls their concerns at the time that the paper would be interpreted in the wrong way. In 2005, as today, there was growing resistance to the theory of evolution, with advocates of intelligent design calling for biology teachers to teach their religious alternative—the idea that

God guides evolution, rather than it being a blind and random process—in class. "When we wrote the paper we were thinking, 'Oh, no, some crazy naive creationist is going to seize on this, and talk about how even the physicists are open to the idea that there is a creator,'" he says, laughing sheepishly. The notion that an alien in a lab built the universe stands apart from the need for a supernatural God or gods to have made it, he says.

So what did Zee ask him to censor in the paper about the implications of a potential discovery? Hsu laughs uproariously as I tell him that Zee now regrets being so heavy-handed, and he recalls it was something along the lines of "The only reasonable conclusion would be that this was a deliberate message and some entity actually had control over the fundamental interactions of our universe, and designed it in a certain way, and cared enough to encode a message that potentially all inhabitants of the universe could receive." He adds with a chuckle, "So, you know, it's a little bit like discovering the tablets of stone with the Ten Commandments."

It's difficult to evaluate the odds that we would ever find such signs—a structured message hidden in the seemingly random temperature ripples already detected. It would involve our universe having been made in that way and the creator choosing to convey this sort of message by these exact means. Hsu emphasizes that this is all highly speculative, but adds the optimistic note that cosmologists are currently scouring the measurements of the temperature differences in the cosmic microwave background in an effort to map it completely, to try to understand more about how the early universe was created. If there are any unexpected regularities—any slight hints of a coded communication—then people will spot them and try to analyze them in depth.

"If you think about stepping outside of theoretical physics for a minute, you just say, 'What's the most exciting thing you could discover?' Well, if someone could give you, quote, 'proof that our universe was created by a loving God,' that would be pretty good, right? And this is coming up with a quasi-realistic way in which that could actually happen," Hsu says.

It's an exciting prospect, yet I wonder whether finding out that we were created, by aliens or by God, would (or should) change our view about becoming creators ourselves. Pain comes as part and parcel of this world, alongside hope and joy. The problem of evil is one of the biggest arguments leveled against believers: how could a loving God stand back and watch His creations suffer? Would we want to be responsible for inflicting life on others?

"I think there are real moral questions that we have to confront," says Hsu thoughtfully. It is something he has been considering over the past decade, not just in the context of creating baby universes but also in terms of developing artificial intelligence (AI) and simulating consciousness in a computer. I had been thinking that those of a religious disposition might be more concerned than nonbelievers about the almost flippant chatter about creating life in a baby universe. But Hsu says something that makes me reassess this arrogant assumption. He notes that if you take a hard-line materialist view of the self and consciousness as emerging inevitably from the right ingredients—without any need for a God to breathe life into a being, or imbue it with a soul—then you must accept that any beings we create that display signs of consciousness and intelligence have just as much right to be considered living beings as we do.

"And any being—whether it's an alien, silicon-based form of life, or it's a dolphin, or something living in my computer—if I become convinced that it's really self-aware and it has feelings and self-preservation motives, I have to really take into account some ethical questions there," says Hsu. (I'll return to this question about whether we live in a simulation, and the ethics involved in creating simulated life—or beings born within a baby universe of our making—in Chapter 10.) The problem, then, is even sharper for atheists because a homemade universe really could evolve life as we know it. "Eventually we'll hit those problems," says Hsu.

✳

After speaking with Hsu, I feel reenergized. He takes the universe-building project seriously but is also aware that it raises moral issues. And he has genuinely excited me about the prospect that if there is a hidden message from our makers in the sky, someone will find it as they scour through measurements of the cosmic microwave background.

There is one possible monkey wrench in the works that could stop astronomers from detecting such a message even if it is there, an obstacle Hsu had not considered but which cosmologist Eugene Lim, at King's College London, noted in a paper in 2015.[11] The idea was put to him at a meeting in Vieques, Puerto Rico, run by the organization that I work for, the Foundational Questions Institute. Lim recalls relaxing on the beach during a break in the conference with physicist David Tong, of Cambridge University. Maybe it was the heat or the heady sea air, but Tong looked to the sky and mused, with his tongue firmly in cheek, "What if God has hidden a message in the sky, but we're destroying it by looking at it?"

Tong was referring to the possibility that God's message—or any information about the birth of the universe—is encrypted in a quantum code. Quantum laws govern the microscopic world and are notoriously strange. The act of observing a quantum object can change its properties, leading Lim and Tong to begin pondering whether astronomers' attempts to read the sky could in fact be tainting any messages they found.

In Chapter 2 I'll delve deeper into the shifting sands of the quantum world. Its peculiar laws, which allow particles to be in two places at once or even pop into existence out of seemingly empty space, play an integral role in explaining how physicists might be able to conjure an entire universe in the lab from almost nothing. Along the way, I'll meet those who argue that the uncertainty inherent in the quantum realm may also give humans free will—and, even more controversially, that quantum experiments have already revealed a hidden arena lying beyond space and time, whence God could intervene in the natural world.

2

Beyond Space and Time: The Quantum Realm

I'm standing on a noisy street in Barcelona, Spain, staring up at the image of an emaciated Christ on the cross. Next to me is Antoine Suarez, a tall, gaunt physicist and psychologist in his early seventies, with salt-and-pepper hair adorning his olive-skinned face. He is talking me through the stations of the cross—from the Last Supper through the death and resurrection of Jesus—carved on the Passion facade of la Sagrada Família across the street.

The church is one of the city's most popular tourist attractions, but it has a deeper meaning for Suarez. He is a devout Roman Catholic, and for him, his own religious trials go hand in hand with his work studying the mysterious quantum laws that govern the behavior of atoms and subatomic particles, as he later explains to me. It is part of his life's philosophy to seek signs of God's hand, not only in tales of ancient miracles we never personally experience but also in our everyday routines. "When you are struggling to solve a mathematical problem, for me, actually, I interpret that as carrying out a conversation with God," says Suarez. "When you get the result, it is that God gives you the gift of your result, for your struggle."

It's a sentiment that resonates with me, though I can also sense it making less religiously inclined readers cringe. I think Suarez can too, because he laughs, puncturing any air of sanctimoniousness, and adds that his students have japed with him that they must be somehow constantly angering

God, because He refuses to answer their prayers and solve their homework problems for them.

Over the course of this book, we'll encounter the vital role that quantum theory plays in building a universe. Quantum laws will be invoked in the coming pages to explain how space and time themselves might have been generated and how our cosmos could have popped into being. Then, crucially, physicists will enlist the help of quantum processes to design a mechanism to recreate this universe-making feat in the lab. You could be forgiven for thinking that quantum theory is some form of magic that allows cosmologists to conjure up anything they wish, with little regard for logic.

So in this chapter I want to uncover how this cosmic magic wand wields so much power. It can transform particles into waves and back, allow objects to be in many places at once, and potentially conjure parallel universes into being. But despite the seemingly godlike abilities it can confer on physicists, the quantum spell book is purely scientific and constrained by strict—albeit weird and wonderful—rules.

Suarez seems like an ideal candidate to help me do this—although his path to becoming a quantum physicist took an unusual route via cognitive psychology. Now based at the Swiss Federal Institute of Technology (ETH), in Zurich, his quest began as much with the hope of understanding the workings of the human mind as with a desire to uncover the foundations of reality. With these goals, in the late 1990s Suarez designed a groundbreaking experiment that combines quantum oddness with the relativistic effects we met in Chapter 1, which were first identified by Einstein. The results of these tests, carried out in 2001, have revealed a hidden layer of reality that, Suarez asserts, lies beyond space and time. It's into this atemporal, immaterial abyss that budding universe builders will have to somehow reach if they want to carve their own baby cosmos.

Suarez's story begins with a confession. Science is meant to be wholly objective. It is fine for physicists to hold their own religious beliefs in private (or, at least, it's tolerated). But pretty much everyone agrees, whatever their religious (or non-religious) bent, that any preconceptions must be parked outside the lab. Your faith should not affect your choice of what experiments to run or how to interpret the results. But Suarez's fascination with quantum theory is embroiled with his Catholicism in a complex manner. While he intellectually separates the two, he also freely admits that his opinions on faith and science have influenced each other over the years and that both have evolved as a result.

Born in the small Spanish village of Allariz in 1945, Suarez was first attracted to science in high school because of the clarity it offered. As a boy he marveled that rudimentary mathematical equations enabled him to calculate the exact trajectory that a ball would take when pitched. Science offered a certain path toward knowledge about the physical world, just as religion offered a definite route to spiritual insight. Rather than being in conflict, they seemed to complement each other perfectly. "I was always asking: is it possible to harmonize what I know through faith and what I know through science?" says Suarez.

This cloistered view of the world was upended during Suarez's undergraduate years, first at the University of Barcelona and then the University of Zaragoza, in the 1960s. General Francisco Franco's authoritarian regime was well established and much hated, and in 1965 a series of student protests erupted in opposition to his policies. "I was involved in the fight for democracy to overcome Franco's dictatorship," Suarez recalls.

Meanwhile, in the classroom, Suarez was introduced to a branch of physics that was just as destabilizing, threatening the security that had been provided by the classical physics of the nineteenth century—the kind that allowed you to precisely work out where a ball would land after it was thrown. That challenge came from the uncertainty of a twentieth-century invention: quantum mechanics. By the 1920s, European physicists investigating the behavior of atoms and light were starting to paint a surreal picture of the microscopic world, replete with fuzzy lines, blurred edges, and hazy, unidentifiable images.

One of the earliest hints that microscopic entities have their own rule book had been picked up by Einstein, who realized that light isn't simply a wave but is made up of particles that pack a different punch depending on their colors. The giveaway was that when you shine blue light onto a piece of metal, the energy from that light is strong enough to knock a few of the metal's electrons from the metal's surface. But when you switch the blue light to red, no electrons are liberated. This is the photoelectric effect, and the cause of this discrepancy between colors was at first puzzling.

Physicists immediately thought to see if ramping up the intensity of red light ameliorated the difference seen with different colors. Their reasoning was that if light is a wave, then by shining more red light onto the surface, the energy would gradually build up, and electrons would slowly but surely start to fly out. But this didn't happen. Einstein figured out that thinking of light as a stream of particles, or bundles of energy, called photons, could explain why that did not happen, solving the conundrum. He argued that the energy of an individual photon is set by its color—and blue photons have more energy than their red counterparts. That meant that while an individual electron on the metal's surface could be booted out of the metal if it was struck by a single blue photon, a red photon would be too weak to dislodge it.

At the microscopic level, Einstein continued, energy is peculiarly rationed into lumps and dished out, like money, in units of currency. (Here Einstein was building on ideas first formulated by German physicist Max Planck.) Today, Barcelona's shops price items in euros and cents, with the smallest monetary denomination being 1 cent. A ticket for a tour of la Sagrada Família costs €15.00, and it would make no sense for the guide to charge €15.005, because in the European Union money simply can't take the value of €0.005, or 0.5 cent. Similarly, the smallest denomination that the energy of tiny particles can take is known as a quantum of energy. A photon can have 1 quantum of energy, or 2 quanta, or 72 quanta, or any other whole-number multiple . . . but not ¾ of a quantum, or 42 ⅓ quanta of energy, say. These insights earned Planck the Nobel Prize in Physics in 1918 and Einstein the same award in 1921, and gave the burgeoning new field of quantum physics its name.

But while pegging light as a stream of photons may have explained the photoelectric effect, it also opened the door to a bunch of new paradoxes about the subatomic world. The trouble was that light still continued to be stubbornly wavy at times when it really should have been more, well, particle-y. Suarez, like many students before him and since, first encountered light's schizophrenic wave/particle identity as an undergraduate, in the form of the double-slit experiment.

It's easiest to picture how the double-slit experiment works for light by first thinking about how water waves would fare in the same test. Imagine a river blocked off by a dam that has two narrow vertical slits cut into it, about six feet apart. As water flows slowly toward the dam and through the two slits, you can imagine two semicircular ripple patterns emanating from the slits, side-by-side, on the far side of the barrier. As these ripples spread out, the two semicircular sets begin to overlap and combine. At some points, two peaks from the two interfering ripples will come together to create an even higher point in the water. At others, a peak will meet a trough and the water will level out. What you'll be left with is a rather pretty dappled water pattern, created by the two sets of interfering ripples.

This is an interference pattern and is a characteristic that emerges when waves combine.

There's nothing strange about this ripple pattern in water; you can recreate it in your own bathtub with a homemade barrier, if you wish. In 1801—more than a century before quantum mechanics appeared on the horizon—English polymath Thomas Young carried out the same test by shining light at a barrier with two slits, and seemed to successfully demonstrate that light is a wave, by producing a similar interference pattern. Today that experiment is conducted using a laser as the light source, with a screen placed on the far side of the barrier. When light passes through both slits, it ripples out, just like the water waves, and those ripples interfere. The far screen is illuminated by a series of bright and dark vertical stripes that correspond to the peaks and troughs in the interference pattern created. These characteristic fringes show that the light is behaving as a wave.

But post-Einstein, physicists knew that this light "wave" is actually a beam of photons. With that in mind, they added an extra twist to the double-slit test to reveal the particle side of light's personality. The advantage of using lasers is that you can turn their intensity down until they spit out only one photon—one particle of light—at a time. Physicists assumed that when you get down to this quantum level, any apparent waviness in the light should vanish. After all, photons are single particles, and so firing one photon at a double-slit barrier is the equivalent of shooting an individual paint ball at a fence with two vertical gaps cut out of it and with a white wall lying beyond it. As Suarez had learned in high school, paint balls are easy to track because they take definite paths, so if you know how you fired them, you can work out where they will land. If your aim is bad, your paint ball will likely hit the fence and never get through to the white wall behind it. But if a single ball does manage to splatter the wall behind with paint, then you can safely assume that it must have traveled through either the left gap in the fence or the gap on the right.

You can go on to imagine firing a series of paint balls at this same fence. You'd expect that those balls that pass through and splatter the white wall would leave two thickly smeared paint stripes on the wall, which line up pretty much exactly with the two gaps in the fence. Similarly, if you carried out the double-slit test with a laser that shoots photons one at a time, you would expect to see two thick stripes of light slowly build up on the screen beyond, lined up directly behind the two slits in the barrier. You would not expect to see a fringe pattern on the screen because, unlike a wave, a single photon should not be able to pass through two slits at the same time, creating interference.

But that seemingly impossible outcome is exactly what Suarez saw in his undergraduate physics class at Zaragoza. It's an experiment that's been repeated time and again in labs around the world and is just as startling each time: when a laser fires a series of single photons at the double slits, an interference pattern of light and dark fringes gradually builds up on the screen, reminiscent of that created by rippling waves. Somehow, individual photons appear to have the ability to pass through both slits at the same time and deposit themselves on the appropriate place on the far screen to mimic a wave.

Things get even odder when you try to scrutinize exactly what's happening here. How can a single photon—a tiny ball of energy—pass through both slits at once? If you place photon detectors at each slit, to monitor whether a single photon really does traverse both simultaneously, the interference pattern immediately vanishes. Instead, when the photons' journeys are being spied on in this way, the photons behave like good little paint balls passing through either one gap or the other in the fence, but never both. This time no interference pattern of bright and dark fringes builds up on the screen; rather, you see two thick stripes of light that line up with the positions of the slits. Light's wavelike behavior collapses.

There are a couple of take-home messages from this. The first is that as long as you don't try to peer too closely at what light—or indeed any other quantum particle—is doing, then it can perform some pretty amazing tricks. Its wavelike nature allows it to be spread across multiple locations, traveling along all possible paths to its end point. But the second important thing to note is that quantum objects can perform the feat of being everywhere at once only when nobody is looking. As soon as you try to catch them in the act, their quantum spell is broken, and the particles settle into one location and choose to take only one path. And, crucially, there's no way to predict with certainty which location or which path of the two the photon will snap into when you do look—it's a choice that's seemingly made on a whim by the particle at the instant it is observed.

Harking back to Chapter 1, it's this fact that observing a quantum system causes it to suddenly snap from an entity of multiplicities to a single identity, changing its properties in the process, that vexed Eugene Lim and David Tong. The cosmologists had been pondering decoding any signatures left in the cosmic microwave background radiation, the photon afterglow of the big bang, that might reveal how the cosmos began. If that information is encoded in the quantum properties of photons in that radiation, then just by looking at it—by detecting that light in telescopes— astronomers would destroy the code.[1]

In 1926, Austrian physicist Erwin Schrödinger encapsulated the uncertainty of the quantum world in an equation. While many equations in physics—such as those that describe the classical laws of motion, allowing young Suarez to track the path of a paint ball—can tell you what will happen at the end of an experiment, Schrödinger's equation is probabilistic. The particle's wavelike nature before it is observed is represented mathematically in the equation by a quantity known as a wavefunction, which enables you to calculate the odds of finding the particle in one definite location, with certain properties, when it is observed, or in another. (In Chapter 4 we'll see that cosmologists have attempted to define a wavefunction for the entire universe, in an effort to explain how it evolves.)

The fact that Schrödinger's equation deals in possibilities rather than in certainties may not sound too impressive. Anyone who has lost a bet or been caught out in the rain after trusting a bad weather forecast may be skeptical of the power of a science that can offer you only probabilistic predictions of what might occur. But there are a couple of important differences between Schrödinger and a bookmaker or a mediocre meteorologist. The first is that the uncertainty in a weather prediction is often a result of the lack of information about the state of the world; if forecasters had more data, they could make a better guess about the future. But the uncertainty in quantum theory, by contrast, is believed by most physicists to be inherent. No amount of extra data will help quantum physicists to make definite predictions. At least, this is according to the mainstream, standard version of quantum mechanics, called the "Copenhagen interpretation" after the city in which it was articulated by one of the founding fathers of the theory, Danish physicist Niels Bohr.

That does not mean that quantum theory is bad. Although you cannot use Schrödinger's equation to tell you exactly what will happen in any one experiment, if you repeat the same experiment thousands of times, you will see that you get precisely the frequency of various outcomes that the equation predicts. For instance, if you monitor a large group of radioactive atoms in the lab and make a note of when they decay (a quantum process), you'll find that though these times vary from atom to atom, the distribution of decay times—the fraction of atoms that decayed after one hour compared with the fraction that decayed after one day, and after one week, and so on—fits the predictions of Schrödinger's equation perfectly.

An unobserved particle's ability to smear out its location and traverse multiple roads simultaneously leads to another handy effect that will pique the interest of budding universe makers. It tells you that there is always some chance, even if only a very tiny one, that when you do look at your atom, your electron, or your photon, you could find it somewhere totally unexpected. The electrons buzzing around in your phone are overwhelmingly more likely to be found zooming round the circuitry inside your handset than, say, exploring Jupiter. But quantum theory says that there is a vanishingly small possibility that one of the electrons might suddenly find itself orbiting the giant planet and wondering how it got there.

The game of chance that can zip objects to surprising destinations, where by rights they really shouldn't be, is called quantum tunneling. One type of radioactivity, in which the radioactive parent atom decays by spitting out a helium nucleus, depends on it, in fact. The radioactive parent atoms cling to their constituent pieces tightly, and if only classical non-quantum laws were at play, then they would hold on to them forever. The force with which they bind their parts together is just too strong to be overcome spontaneously. But quantum theory tells you that there is a slight probability that a helium chip off the radioactive block might suddenly find itself splitting off from its parent and leaving home—even though, classically, it should not have the energy it needs to do so. Because it might happen, in the quantum world it will happen if you wait long enough. Taking this to the limit, in Chapter 5 we'll see that cosmologists have attempted to explain how a whole universe might be born from (seemingly) nothing by quantum-tunneling its way out of the void and into existence.

Suarez was learning about these strange ideas as an undergraduate, but at that stage he had other, more pressing thoughts on his mind than the foundations of physical reality. The student uprising against Franco had failed. Its previously moderate leaders began to embrace increasingly left-wing ideologies and became more radicalized—a situation that did not sit comfortably with Suarez. Unhappy now with both the Spanish government and the protest movement, Suarez decided to relocate to Switzerland and start afresh. There he undertook a master's in experimental physics at the University of Fribourg, but it was questions of human nature and how children learn that really fascinated him. Why do some infants grow up and choose to support democracy while their friends choose to become dictators? So Suarez took a detour in the 1970s, earning a PhD on cognitive development in children at ETH, jointly supervised by psychologist Hardi Fischer and mathematical physicist Res Jost.

The question of whether science allows humans to really have free will has always intrigued Suarez, and it brought him back to foundational physics. If our brains are just conglomerations of neurons, firing away according to basic physical laws, then it could be argued that we have little to no control over our actions. We may think we are making free choices based on our wants and needs, but perhaps those desires were just the inevitable outcomes of a chain of chemical reactions in our brains that were determined at our birth. Winding the clock back, these, in turn, were causally linked to events in the universe's history, going all the way to the big bang. As French mathematician Pierre-Simon Laplace had stated it in the nineteenth century, a superior intelligence, or an all-knowing demon, who knew the position of every molecule in the cosmos and had a complete knowledge of science could calculate the fate of everyone and everything to come.

While he was at ETH, there were two triggers that set Suarez pondering whether the quantum theory he had learned twenty years earlier had interesting implications for the issue of human choice. He struck up a long-term friendship with mathematician Ernst Specker, who, with his colleague Simon Kochen at Princeton University, had derived a seminal theorem about indeterminism in quantum theory in the 1960s.[2] Specker, who passed away in 2011, was also an evangelical Christian and preached at a church in Zurich on Sundays. Suarez and Specker discussed the differences between their branches of Christianity. Suarez laughs as he recalls that Specker once asked him whether, according to Catholic teaching, you can repent for one sin but not for all—although Suarez never knew which particular sin was burdening Specker—and also whether he thought that physics had room for angels.

Together, Suarez and Specker also wrestled with the issue of free will, but with a quantum spin. Since the standard interpretation of quantum theory says that, deep down, nature is indeterministic, then this might break the chain of deterministic chemical reactions from the big bang to today that otherwise appears to shackle our fates. Perhaps, they mused, quantum theory opened a big enough chink in determinism to allow humans to make free choices.

But there was a downside. Specker was deeply troubled by the theological dimension of the debate. First there was the centuries-old religious puzzle of whether humans can really have free will, if their fates are predestined and known by God. Quantum theory only complicated things further. Specker was devoted to the notion of divine omniscience, and while quantum indeterminism provided a glimmer of hope that humans might have free will, at the same time it seemed to rob God of His all-knowing power. According to standard quantum theory, prior to observation, the characteristics of quantum particles are undefined and the outcome of measuring them is inherently unpredictable. This cast a veil over the deepest layer of reality, even calling into question whether it made sense to say that there was a fuzzy layer of reality there at all before you looked. What unsettled Specker, and then later Suarez, was that potentially even a deity could not penetrate this veil.

There are, however, nonstandard interpretations of quantum mechanics that retain determinism, and the duo discussed whether these might help to reconcile their faith with quantum theory. One of the most exotic deterministic alternatives to conventional quantum theory is the many-worlds interpretation, proposed by American physicist Hugh Everett III in the 1950s. It conjectures that our universe is just one of a number of parallel cosmoses. Each time a measurement is made in a quantum experiment that has two or more possible results, reality branches, creating parallel worlds—one for each possible result that could be realized. According to Everett's speculation, when Suarez set up his double-slit experiment as an undergraduate, the moment he made a quantum measurement of the path of the photon—to check if it took the left slit or the right slit—reality fractured, creating two almost identical clone universes. The only difference between the two would be that in one the young Suarez detected that the photon traveled through the left slit, while at that same instant his parallel self in an alternative version of Zaragoza was recording the outcome that the photon had traveled through the right slit.

This sprawling multiverse of infinitely branching universes with multiple selves had the advantage that reality was now no longer a game of chance. Rather, everything that could happen did happen. Although it was fiercely deterministic, it did not appeal to Suarez's sensibilities. "For me, many-worlds removes free will and it removes moral accountability," says Suarez. After all, what's the point in choosing to behave well and work hard in the lab when another version of yourself will inevitably choose to skip work, while a third will end up robbing a bank? Okay, God may know everything that happens in the quantum multiverse, but He couldn't very well judge you on what you did since, by necessity, some version of you had to do bad things. (In Chapter 9, we will meet a physicist who takes the contrary position that the many-worlds view does allow for moral responsibility and can be reconciled with Christian faith.)

In addition to his chats with Specker, something else happened that pushed Suarez to study quantum theory more deeply. In 1988, Suarez read a review of the book *Speakable and Unspeakable in Quantum Mechanics*, by Irish physicist John Stewart Bell, in *New Scientist* magazine. The book described a bizarre quantum property called entanglement that appeared to allow for instantaneous, faster-than-light communications between distant particles. It was a feature that had niggled at Einstein, who denounced it as "spooky action at a distance" because it seemed to clash with the speed limit laid down by his own theory of relativity.

Bell himself had in the 1960s devised a way to test whether entanglement did allow such a ghostly link between particles. It took a while for technology to catch up with his designs, but by the late 1980s, when Suarez was reading his book, those experiments were being carried out with ever greater precision and appeared to be establishing that entanglement was real and just as spooky as Einstein had feared. Suarez read Bell's words and was spellbound. "I said to myself, 'I have to meet this guy,'" says Suarez.

Suarez called up Bell, who was based at the CERN laboratory near Geneva, Switzerland, out of the blue and was surprised that Bell not only was friendly enough to talk to this stranger but invited him over. And in the flesh, Bell was just as mesmerizing on the subject of quantum experiments and their interplay with philosophy. "After speaking with him, I made the resolution to dedicate myself to this subject until the end of my life," says Suarez. Bell became his friend and mentor until his untimely death of a brain aneurysm at sixty-two, in 1990, and Suarez's voice still trembles when speaking of the shock of this loss.

Bell's legacy was to inspire Suarez to devise his own experiment focused on entanglement—the test that would ultimately reveal a layer of reality that lies beyond space and time. The word "entanglement" has a romantic ring to it, and one way to understand why it is so disconcerting is to first imagine a classical analogy involving two lovers, Alice and Bob, who are about to part for some months. They have a love token, a yin and yang pendant, which they split apart into a white piece and a black piece. They place the two halves in two identical envelopes, then seal and shuffle the envelopes. Alice takes one envelope with her on her journey to Spain, as a memento, while Bob takes the other and heads to Switzerland. When they arrive at their destinations, missing each other, they open their envelopes and look at their half of the pendant. Bob, seeing that he is holding the black segment in his hand, instantly knows that Alice must possess the white fragment. Alice, meanwhile, looks at her white portion and similarly concludes that Bob's is black.

There is absolutely nothing perplexing about that macroscopic pendant story. But in 1935 Einstein, along with colleagues Boris Podolsky and Nathan Rosen, published a thought experiment outlining the equivalent process in the quantum world and revealed a paradox. In the quantum version, the two halves of the pendant are replaced by two particles. For instance, in a superconductor, electrons tend to naturally hang around with each other in pairs, with complementary properties, so that if one of these twinned particles is spinning in a clockwise direction, its entangled partner will be spinning in a counterclockwise sense to balance it out.[3]

Again, as with the pendant example, one electron is sent to a physicist called Alice, who is waiting in a lab in Spain to measure the sense in which it is spinning, while the other is sent to her colleague Bob in Switzerland, who is poised to observe his electron. Their measurements are the equivalent of both parties opening their envelopes. When Alice sees that her particle is spinning in a clockwise direction, she immediately knows that Bob's particle must be spinning counterclockwise, without her having to ask him. Entanglement means that the properties of the two particles are forever intertwined. Likewise, after Bob studies his electron and sees that it is spinning in a counterclockwise direction, he can be certain that Alice's twin particle is spinning clockwise.

At first that may sound no stranger than the case of the yin and yang pendant halves. But here's where quantum wizardry comes in. The double-slit experiment and many others since tell us that until the moment when Alice observes her particle, the properties of that particle are indefinite. It's as though she was holding a closed envelope containing a pendant piece that is both black and white, as well as every shade of gray in between, and it's only upon tearing open the envelope that the pendant fragment's color is set to either black or white. Prior to measuring the particle, the odds of Alice measuring spin in either direction are equal. The same is true for Bob; before either he or Alice looks, the odds of him measuring either a clockwise or counterclockwise spin are likewise 50/50.

But at the instant that Alice finds that her electron has settled into a clockwise spin, say, the rules of entanglement insist that Bob's particle will definitely have a counterclockwise spin when he measures it. Had the odds fallen the other way and Alice observed the spin of her particle to be counterclockwise, she would know that Bob's particle will definitely be spinning clockwise when he observes it.

This is the part that galled Einstein. Alice's actions in Spain appear to instantaneously fix the fate of Bob's particle in Switzerland—despite the fact that the two electrons are hundreds of miles apart. In Chapter 1, we saw that Einstein's theory of relativity had set an upper limit on the speed that light, or indeed anything else, including information and communications, can travel. Quantum theory, however, seemed to suggest that an experiment carried out in one part of the world can magically influence another in a distant country—or even, in principle, in a different galaxy— even though there should not be enough time for any kind of signal to travel from one lab to the other, without breaking relativity's speed limit.

Little wonder that Einstein regarded entanglement as spooky. This clash with relativity convinced him that quantum theory was incomplete—just an approximation to a deeper, deterministic framework of reality in which particles had set properties, even while nobody was looking at them, and in which entanglement's correlations could be explained without resorting to ghostly faster-than-light influences. But experiment after experiment seemed to suggest that Einstein was wrong: the properties of twinned sets of particles kept on eerily matching up. These "correlations cry out for an explanation," Bell had once said—and Suarez was resolved to find it.

Suarez, Bell, and others were aware that these entanglement experiments that purported to show spooky correlations were not watertight. One important loophole lay in the fact that Alice and Bob were not actually located in different countries but were often only in different parts of the same lab. This meant that the experimenters could not truly rule out the possibility that when Alice made her observation, she somehow triggered a signal from her particle that was relayed through the apparatus to Bob's particle and influenced the result of Bob's measurement. If there was even a split-second time difference between Alice and Bob's measurements, then that communication could have traveled at less than the speed of light—which would mean that quantum theory was perfectly in line with Einstein's theories and not spooky at all.

Suarez decided to design a stricter version of the entanglement experiment that broke off any possible line of communication between Alice's and Bob's particles. His primary reason was purely scientific: like Bell, he had a burning desire to know what was happening at the fundamental level, and he knew that technology had now reached the point where physicists could find out. "But I was also fascinated by Specker's theological motivation about the relationship between divine omniscience when tackling foundational questions of quantum mechanics," he adds. Could he prove, as Specker had hoped, that quantum mechanics was wrong and would yield to a more fundamental deterministic framework that enabled God to know all that is happening? Says Suarez, "For me, it was always a background motivation."

Like Einstein, Suarez had a gut feeling that the weirdness of quantum theory was just an illusion created because we don't yet know the true, more commonsense description of reality. So when designing his entanglement experiment, Suarez was convinced that these particles could not really have any spooky signaling powers allowing them to quickly tell each other, at faster-than-light speeds, to match up their properties. He was certain that if he succeeded in stopping any possibility of a communication traveling at less than light speed between the two particles, then the correlations between their properties would vanish. He and his colleagues would thus prove that quantum theory was far more mundane than physicists had been led to believe.

To jam any signal between the particles, Suarez employed a neat trick, with the help of Einstein's special theory of relativity. He recalled that—as we saw in Chapter 1—relativity can tinker with the ordering of time itself. In particular, relativity tells us that when two people are in motion relative to each other, they can disagree on which of two events happened first. Going back to the pendant case, imagine that Alice and Bob were actually both astronauts, boarding two spaceships zooming off in opposite directions. Then, according to relativity, it is possible to set up a scenario in which, from Alice's point of view, she opens her envelope before Bob, while from Bob's perspective, he tears his envelope before Alice does. Both would be correct from their own perspective, because time ordering depends on motion.

If Suarez could recreate those conditions in the lab—without spaceships, preferably—then it would shut down the possibility that if Alice (according to her point of view) measured her particle a split second before

Bob, a signal could be transmitted from Alice's particle to Bob's instructing it on how to fix its own properties. That's because from Bob's perspective, he observes his particle before Alice, and so any communication would need to travel in the other direction. "It would not be possible to explain the correlations by saying that an act of measurement made on the first particle influenced its partner, because, depending on how you looked at it, both particles could be said to have been measured before the other," says Suarez.

Suarez developed his aptly named "before-before" experiment in 1997 by modifying a standard entanglement test in conjunction with his colleague Valerio Scarani, who is now a quantum physicist at the National University of Singapore.[4] In the normal version of the test, which had already been carried out many times by independent physicists, two entangled photons, A and B, are created with a laser beam. Each photon follows its own path around a table before hitting a beam splitter—a half-silvered mirror that serves as a crossroads, beyond which the photon can take either a long path or a short path. The odds of a photon taking either path are 50/50, but thanks to their entanglement, in every trial of the experiment whenever photon A took the long path, so would B; similarly, if the first to hit the beam splitter took the short path, so would its partner.

To bring relativity into the mix, Suarez and Scarani needed to house the two photons on the equivalent of two mini spaceships hurtling in opposite directions in the lab. They did this effectively by using two moving beam splitters.[5] Because of the relative motion between the two beam splitters, as each photon hit its own beam splitter, an imaginary miniaturized Alice running alongside photon A would swear that her photon hit the beam splitter and chose its path before photon B, while a tiny version of Bob would likewise be certain that his photon B had hit the junction and picked a trajectory before photon A.

The before-before experiment was carried out in 2001 at a lab run by Nicolas Gisin at the University of Geneva, Switzerland.[6] Suarez recalls their eager anticipation of the results. He was certain that monkeying around with any possible cause-and-effect relationship between photons A and B would destroy any quantum correlations. If so, entanglement should vanish, and he and his team would have shown that Einstein was right to distrust the apparently magical powers of quantum theory and advocate a search for a sensible, deterministic theory of reality in its place. "All the team was awaiting results that would refute quantum mechanics and get us the Nobel Prize," says Suarez. They were so convinced that they would upend quantum theory that they discussed the importance of checking and rechecking their results before going public, as the conclusions were bound to be groundbreaking.

And indeed, their result was monumental for Suarez, though it did not "beat quantum theory," as he had hoped. Instead, on every run of the experiment, the photon pairs' fates continued to match up. The team had not only strengthened the evidence for spooky action at a distance but also shown that quantum theory exerted powers greater than anyone had yet imagined. Somehow the correlations between particles endured even though the team had broken the possibility that there might be a temporal link between the measurements. "The experiment led me to the enlightenment that these correlations are timeless," says Suarez.

What does that even mean? From a philosophical point of view, Suarez argues, it reveals the existence of a fundamental realm—a layer of reality that controls the quantum behavior of particles—beyond space and time. "To explain the visible world you need to accept influences coming from the invisible," says Suarez. "You cannot explain these correlations by any story registered in space and time." Quantum theory had bested Suarez and defeated space and time themselves.

While we are usually taught to keep science and religion separate, Suarez is someone who was driven by religious considerations to scrutinize physics in a more rigorous way. I should add here that his colleagues did not necessarily share his faith and were driven more by an aesthetic dislike of some of quantum theory's spooky attributes rather than by worries about God's limits. Crucially, however, when the results the team found did not fit with Suarez's preconceived prejudices, he threw out his misconceptions and embraced quantum theory in all its indeterministic oddity.

There are philosophical pros and cons in accepting quantum indeterminism, Suarez notes, and even today, he is still wrestling with its implications for his faith and his understanding of humanity. For religious reasons, he has always believed that humans must possess free will, in order to be accountable to God for their choices. Suarez now firmly believes that human free will is intimately tied to quantum indeterminism. He argues that it is necessary to invoke quantum unpredictability to explain how humans can break the cause-and-effect chain governing the molecules that make up our brains that would otherwise predetermine the choices we make. This is by no means a mainstream stance, however. While quantum theory can provide the "freedom" in free will, critics note that it doesn't explain the mechanism through which our consciousness might control our behavior. Suarez and a small cadre of philosophers who share his views must provide an explanation for how the "will" part of free will arises.

And what of God? Now that Suarez has accepted that reality is indeterministic, can he make peace with the notion that maybe not even God knows the current state of reality? It's a problem that has never ceased to trouble him, and in recent years he has found himself reexamining Everett's many-worlds hypothesis, which asserts that every possible fate allowed in a quantum experiment is deterministically realized, each in a separate universe. Though he still rejects the notion that such parallel universes are physically realized, he is working on a compromise viewpoint. "If these parallel worlds are only the mental contents of some mind—the mind of God—then the result is quite appealing," says Suarez. In this picture, God's omniscience manifests because He knows all possible fates and the consequences of all possible choices and all the possible worlds they create. "What your free choice does is to fix which one of these possible worlds is physically realized," says Suarez.

Suarez has now walked a good few theological steps further than most of his colleagues would feel comfortable with, though he acknowledges with a smile that they do listen to his ideas, even if they do not always agree with him. A case in point is his interpretation of the results of the before-before experiment. Most physicists are happy to say that the test has shown that quantum laws play out in an arena outside space and time. But they would stop short of Suarez's next assertion: that this realm could be interpreted as a divine playground. "It is in agreement with the major religious traditions that accept there is an eternal and immaterial God," says Suarez. "The science that we are doing today is in symbiosis with principles which we call 'spiritual.'"

Not everyone will want to follow Suarez's line of reasoning this far. Certainly, recognizing that a layer of reality lurks beyond space and time is not proof of the existence of God—and Suarez does not make that claim. But for those with a religious bent, like him, the scientific discovery resonates with their theology.

Since Suarez's before-before experiment was carried out in the early part of the twenty-first century, many other experiments have continued to find even stronger support for entanglement and other strange facets of quantum theory.[7] For most physicists, the take-home message has been that the world is fundamentally more bizarre than we can easily imagine. Strip away space and time and what you're left with is the quantum.

But the quantum *what*?

There can't be any quantum objects there—no photons, no electrons, nothing with spatial or temporal extent. There isn't even any "there" there.

What you have are quantum rules that apply to what physicists call fields. Even an apparently empty field is bursting with the potential to give rise to a sea of somethings: to matter, to light, and, as we will find in the following chapters, to universes. In Chapter 3, we will see how such brimming quantum fields populated our universe. But before we get to that, to help me visualize the power of fields, I am going to make a brief detour to Pennsylvania, to meet a physicist who is renowned as a deep thinker, both for his contemplations on how space and time arose in the external world, and for his spiritual meditations on the inner world of human consciousness.

＊

My flight into State College, Pennsylvania, would make atheists start to pray for God—first that He should pop into existence and then that He should bring the craft down safely. I am not a great air traveler at the best of times, and my long-haul trip from London to Washington's Dulles Airport had met with turbulence that made for a bumpy landing. So it was with great reticence that I boarded this connecting flight—in a plane that is barely the size of a minibus—for a short hop in those same strong winds to what's effectively the Penn State University airport. When Emily, the flight attendant, informed me and the other three passengers that we would need to vacate our assigned seats at the front of the craft and move back a couple of rows to "balance the weight of the nose," and that we should prepare for an unusually rough journey, my heart sank.

The plane teases me with moments of calm in the air before suddenly lurching downward. "Why're you visiting Penn State?" Emily asks brightly. She's noticed my distress and is trying, in vain, to distract me from this roller-coaster ride with small talk. I am on my way to meet Abhay Ashtekar, a physicist who is one of the originators of a theory that attempts to unravel Einstein's fabric into its constituent quantum threads and answer the question of what came before space and time. It's a model he has developed in tandem with his own meditative explorations of Buddhism and other Eastern philosophies, with interplay between his science and his spirituality,

I say. Or at least that's what I would say if I wasn't too scared to open my mouth in case I lose the semi-digested remains of my in-flight meal from the previous leg of the journey, which are threatening to resurface. So instead I fail to smile back, managing only a pitiful grimace.

As I try to clear my mind of panic, taking long, deep breaths, it strikes me that it's no surprise that someone like Ashtekar, who has to make this journey on a regular basis, should also feel the need to become well versed in the art of meditation. I hazard this theory to Ashtekar when I eventually do land and meet him at the university. He guffaws, and then says in an incredulous voice, "Was the flight that bad? I have made it so many times, I don't even notice it." Perhaps, I think, I am being overly sensitive about the flight.

Ashtekar's interest in meditation, along with his love of physics, has roots that go far further back than his time journeying to and from Penn State, where he now heads up the Institute for Gravitation and the Cosmos. They stretch back to his childhood in India in the fifties. "I guess in India, there's just something in the air," says Ashtekar. "There are just many people interested in big questions."

As a kid, Ashtekar was bitten by the physics bug when he read the popular book *One, Two, Three . . . Infinity* by Russian physicist George Gamow (the same volume that enraptured a young Tony Zee, as we saw in Chapter 1). He acknowledges that it's something of a mystery as to where he got hold of the book. His father was a civil servant, and Ashtekar's early years involved bouncing from one small town to the next, following his father's postings. It was rare to find physics texts in the town in which he was schooled, though they were replete with works on ancient Hindu philosophy. "It was far easier to find a book on Vedantic philosophy than a book by George Gamow," Ashtekar chuckles.

Though he is now in his mid-sixties, Ashtekar's dark brown face, crowned with thick waves of curly black and gray hair, is almost always lit up with boyish animation, and he appears to possess limitless energy. Before we settle down to talk, I attend one of his cosmology seminars, in which he's outlining a calculation he has just cracked on the evolution of the early universe. Ashtekar spends much of his allotted hour running back and forth in front of the blackboard that covers one wall of the seminar room, his arms gesturing wildly as he conveys his enthusiasm at solving the pesky equation that had been perplexing him. Ashtekar is a man who feels he has answers to some of the deepest mysteries of the universe, and he clearly wants to share that joy with the world—starting with his gathered students.

That energy abounds during our talk too, as he is eager to discuss his unconventional theory of the origin of the cosmos, loop quantum gravity. When summarized, it sounds almost embarrassingly trivial. Ashtekar, who is wearing a gray cotton shirt, pulls at its sleeve to illustrate his theory. "If you look at my shirt from far away, it appears to be one smooth, continuous piece of material—a two-dimensional continuum," he says. "But if I just look at it with a magnifying glass, I can see that it is woven together by one-dimensional threads—and the idea is the same thing is true with the fabric of spacetime itself."

It's not quite that straightforward, of course. For starters, there is no magnifying glass powerful enough to reveal these threads of spacetime, which are called loops. Any such instrument would need to focus down to a minuscule distance of little more than a billion of a billionth of a meter, and no such experiment has even been envisaged. Since loops cannot be directly seen, Ashtekar and his collaborators—most notably physicists Carlo Rovelli, at Aix-Marseille University in France, and Lee Smolin, at the Perimeter Institute for Theoretical Physics, in Waterloo, Ontario—have developed an intricate mathematical framework with which to attempt to convince people of their existence and explain facets of cosmic history.

So Ashtekar's threads remain hypothetical entities, loops of energy that he pictures spontaneously manifesting through quantum processes. The fertility of quantum theory that could in principle enable it to manifest these loops lies in its inherent uncertainty. The inability to say with confidence that a particle is in any one place at any one moment is so profound that even in the absence of space and time it is impossible to identify emptiness with absolutely nothing. There's always some probability that the quantum vacuum, apparently void of all matter, will in a split second birth pairs of particles—say, an electron and its antithesis, a positron (a particle with the same mass as an electron but the opposite charge). These pairs usually only taste existence for a fleeting moment before recombining and disappearing back into the quantum ocean again.

Einstein had realized that photons are bundles of energy, and physicists now regard them as excitations—little lumps of coherent energy—in the electromagnetic field that Maxwell identified with light in the nineteenth century. Similarly, Ashtekar and his collaborators argue, the seeds of spacetime are spontaneously created as excitations, or high-energy states, bubbling out of an underlying field of quantum geometry. These are the loops of energy that weave together in their model to knit Einstein's spacetime fabric.

I am not sure how you can have a pervasive field of quantum geometry that exists before space itself is there to manifest that geometry. I put that criticism to Ashtekar, who simply replies that before Einstein, physicists could not conceive of spacetime as a dynamic entity that existed without matter—and yet now we know that spacetime is like a fabric that stretches and warps around matter. "Until Einstein came along, spacetime was an abstract thing, it was like a stage or a canvas on which you painted things, and it was the painting that was real and physical," says Ashtekar. Now, however, we know that spacetime itself is a real physical entity.

To drive that point home, in 2016 physicists working on the Laser Interferometer Gravitational-Wave Observatory (LIGO) experiment announced the first detection of gravitational waves, ripples through spacetime itself that were set in motion when two black holes collided and which were then picked up here on Earth over a billion years later.[8] If spacetime were not a real physical entity, it would not have been able to transmit these ripples. But while gravitational waves are undulations of spacetime, Ashtekar posits that loops are the physical excitations of quantum geometry that give rise to spacetime itself. "Quantum geometry is always there, it is eternal," he says.

Ashtekar is unabashed about the parallels he sees between his physics research and his readings into Eastern philosophies. For instance, musings on the notion of eternity and timelessness are familiar from his readings of the ancient religious texts that suffused his surroundings in childhood. While Hinduism is the majority religion in India, Ashtekar's family was actually part of the minority Jain faith, which rejects the notion of a creator deity and emphasizes nonviolence to humans and animals, along with self-control as a means to enlightenment for one's immortal soul. Though religion did not play a large part in his upbringing at home, he was always fascinated by alternative religious thinkings, such as Hinduism, Buddhism, and Chinese Taoism, that he learned about from the various communities around him as his family hopped from town to town around India.

After studying physics at the Indian Institute of Technology in Mumbai, Ashtekar came to the United States, first to the University of Texas, in Austin, and then to the University of Chicago, where he studied for a PhD under Robert Geroch, an expert on Einstein's general theory of relativity. There he also met astrophysicist Subrahmanyan "Chandra" Chandrasekhar, famed for his work on the evolution of stars, which would later win him the Nobel Prize in Physics in 1983. Both men served as mentors for young Ashtekar. Geroch taught him "mathematical clarity," while Chandra encouraged him to have the confidence to tread an unbeaten path and pursue questions that other physicists were reluctant to address. "He taught me that you have to be brutally honest about doing things that you think are important as opposed to what is fashionable," says Ashtekar.

Ashtekar's other great influence was Roger Penrose, who—as we saw in Chapter 1—had helped to establish that time began at the big bang. Along with Stephen Hawking and George Ellis, he had identified that the universe originated from an infinitesimally small but monumentally dense point called the singularity. (Geroch had also written papers on this subject.) How this singularity arose, and what exactly happened there, remained a mystery, however. Ashtekar joined Penrose at Oxford University for his first job upon graduating from Chicago and credits his then boss with teaching him the "art of dreaming"—allowing his intellect to soar along flights of fancy and having the confidence to see them through no matter where they led. Ironically, when Ashtekar took that advice to heart and developed his own looping model of spacetime, he inadvertently overturned Penrose's assertion that the big bang marked $t = 0$.

It turns out that in a world in which spacetime is sewn together from loops, the loop itself roughly demarcates a minimum size below which spacetime cannot be squeezed. (More accurately, just as energy is quantized in quantum theory, with a certain minimum value of one quantum, area is quantized in loop quantum gravity, so there is a minimum size below which loops cannot thread together to create spacetime.) So in theory, in this model, the universe could never have been squashed into an infinitely small, infinitely dense point, or singularity, even at the time of its birth. To check what loop quantum gravity predicts happened in our infant universe nearing the big bang, Ashtekar's colleagues created a computer simulation of the cosmos as it is today and then rewound time across the eons. They watched the cosmos shrinking down, as expected, as the clock ticked backward, approaching $t = 0$. But then, roughly 13 billion years ago, it reached a small but finite size—the limit set by the quantum geometry.

When the universe hit this minimum size, it started to expand again, apparently pushed outward by a repulsive force. The team interpreted this result as showing that our cosmos had no finite beginning in time, no big bang moment, but has existed eternally.[9] At some point in the past, it contracted, bounced, and then grew again. Our current expanding phase, according to loop quantum gravity, is just the latest stage of this rebounding history.

There are parallels between this picture and that painted by Indian religions, in which time is eternal, without beginning, while the universe cycles through phases of creation and destruction. The resemblance is coincidental, however. Loop quantum gravity was not designed to generate an eternal cosmos; this feature simply fell out from its mathematical equations. Ashtekar was so stunned to discover this bouncing prediction of his theory, around a decade ago, that he initially dismissed it as a programming glitch. "I was very taken aback by this," he laughs sheepishly. "For six months, actually, I didn't let anybody publish this result." Only after numerous checks was he convinced it was real.

But while this resonance between Ashtekar's physics and Eastern philosophy is just a coincidence, his science and spirituality come together in other more overt ways. Over the past fifteen years, Ashtekar has dedicated himself to the Buddhist practice of Vipassana meditation. Buddhism speaks of breaking the cycle of reincarnation by detaching oneself from the physical world and attaining enlightenment, or nirvana.

As he speaks of his spiritual beliefs, Ashtekar's boyish, energetic demeanor gives way to a calmer, more serious manner, and his age appears on his face. He recasts his worldview in terms of physics, using the same language of excitations of quantum fields that he uses to describe his theory of loop quantum gravity. "My viewpoint is basically that there is a field of consciousness," says Ashtekar. Individual human consciousnesses exist as agitations in this ocean of communal consciousness—just as photons are excitations of the electromagnetic field and his hypothetical loops are energetic nuggets of spacetime.

Everyday stresses, such as taking a rough flight into State College or struggling to meet a deadline, Ashtekar asserts, create highly energetic tensions that pull our minds out of the sea of consciousness like turbulent waves. The suffering we experience, says Ashtekar, is a manifestation of the distance between our individual energized state of consciousness and the stable background field. Peace can be achieved by simply sinking back in and unifying with this ocean. "Perhaps nirvana is just the ground energy state"—the state of lowest energy—"of this consciousness field," Ashtekar says.

This is more than a metaphor. For Ashtekar, the consciousness field is a real "physical entity that is all-pervading." He is aware that some may scoff at his conjecture. "Most scientists don't even want to acknowledge that there is an inner world," he says. Ashtekar preempts the criticism that such a field has never been detected by countering that it took a century for physicists to find evidence of the gravitational waves in spacetime that Einstein had predicted in 1916.

But ultimately, the best proof comes through personal experience. "The only way I can convince you is to tell you to do these meditations yourself," says Ashtekar. Adherents of the Vipassana school attend ten-day retreats at which they refrain from speaking, reading, writing, or other stimuli. The process is mentally draining, far outstripping the dedication needed to conduct research in physics. "From my intellectual life I had the inner pride of being able to concentrate for hours when I am working on something," says Ashtekar, noting wryly that he often becomes so absorbed in a calculation that he forgets to eat. But at the retreat, Ashtekar realized that this level of mental exertion pales in comparison to the concentration needed to move one's mind "beyond thought." "They say that the first time that you do it, it is like surgery for the mind, and it does have a very deep cleansing effect on your consciousness," says Ashtekar. "When drained of all sensations and perceptions, there is still this residue—and that is your kernel, the seed inside you of consciousness, interacting with the field of consciousness."

When his mind was quieted after ten days of meditation, Ashtekar recalls, tears ran down his cheeks, "tears of gratitude—to whatever" for the realization of how irrelevant daily tribulations can be and that life "can be so simple and so beautiful." Ashtekar speaks these words with utter sincerity, and it is hard not to be both moved and inspired by them. Some could dismiss his experiences as hokey, but I choose to interpret them as profound.

The effect of deep meditation was so pronounced that Ashtekar actually found himself in danger of becoming too absorbed in his inner world, dulling his research ability. "You lose your motivation, or the fire in your belly, or whatever you need to do something very deep in the external world," he says. It's led him to step back from intense meditation in an effort to remain anchored in the everyday. His scientific research now focuses on examining whether loop quantum gravity makes concrete predictions about the pattern of temperature ripples in the cosmic microwave background—the leftover radiation from the big bang that entranced Zee (see Chapter 1)—which astronomers might be able to detect in order to corroborate his model.[10] But Ashtekar still dips into meditative practice at times of stress. "I use this to have basic peace," he says. "It brings in basic joy."

Ashtekar remarks laughingly that his musings on consciousness are likely to irritate his scientific friends. He recalls trying to share his spiritual quest with Chandra, a vocal atheist, who had suffered a heart attack during Ashtekar's final summer at Chicago. As a gift to read while recovering, Ashtekar gave his mentor the classic sixth-century Chinese text *Tao Te Ching*, by Lao Tsu, a contemplation on the best way to live, according to Taoism. He hoped it would take the senior man's mind away from the anxieties of work and the "nitty-gritty of equations" and instead inspire him with words and concepts that were "soaring high."

Chandra was surprised by the gesture and, upon receiving the book, asked Ashtekar to name his highest goal. When Ashtekar responded that it was "to understand the inner world and how perception occurs," Chandra seemed disappointed. "I think he expected a different answer, one that focused on the external world or the science of the physical universe," Ashtekar recalls. "It's not that he did not respect the spiritual life, but he thought there were just too many charlatans involved with religion, too many people saying things that they did not understand—which is absolutely true." But Ashtekar is still niggled by the feeling that Chandra undervalued examination of the inner world. In doing so, Chandra failed to live up to words he had once told his student when describing how to choose his scientific area of specialization.

"Chandra had told me that when it comes to scientific fields, you should judge them by the work of the best people, not by the work of the mediocre people," explains Ashtekar. He now regrets never having turned to his old mentor and saying directly that he should practice the same standards he preached before dismissing the merits of spirituality out of hand. "There are many charlatans, there are various people saying all kinds of things—completely stupid, cheapening it all—but that's not what you judge it by," says Ashtekar.

As I board the little plane at the State College airport to make the return hop to Dulles, I encounter Emily the flight attendant again. "Oh, hi! I remember you," she says cheerfully. Turning to the other passengers, she points to me and says, "This lady was on the worst flight I've ever flown last week—there was a storm, and the plane was plummeting. I couldn't even stand up." Aha! I feel vindicated that I was not overreacting.

"So, did you get what you came for?" Emily asks. I take a moment to think it over. I have learned about Ashtekar's model for the early universe, but I also know that it is controversial. Though his work is widely respected and his mathematics is sound, it is by no means the only fundamental model of reality on the market. In fact, in Chapter 7, we will meet one of its rivals, string theory, and we'll learn that elements of string theory, if it is correct, could make universe building a whole lot easier.

But Ashtekar has taught me about the basics of quantum fields and how they can conjure particles seemingly from thin air. Those fields will play a crucial role in explaining how our universe inflated after its birth—whether born from a conventional big bang or from an exotic big bounce. And it's this inflationary process that is the core mechanism that needs to be harnessed by physicists to manufacture a universe in the lab. In Chapter 3 we'll meet the founding father of inflation, Alan Guth, one of the first men to consider the possibility of making a homemade cosmos.

I turn to Emily and smile. "No, I haven't quite got everything I need just yet," I say. Now that we've covered the background, the quest to build a universe in the lab begins in earnest.

3

Inflating the Universe

In 2014, cosmologist Alan Guth received one of his strangest requests: a colleague asked him to swear an affidavit stating that his research does not prove that the universe had a beginning.[1] The document was then used as evidence in a debate to argue that there is no need to invoke a God or gods to explain the origin of the cosmos.[2] Guth is best known for having come up with the theory of inflation, an extension to the big bang theory that's now pretty much mainstream, more than three decades earlier. This is the idea that in the moments after the big bang, a tiny pocket of the infant cosmos rapidly expanded to create the huge patch of the universe that we now call home. It's also Guth who first realized, some thirty years ago, that, in theory at least, budding universe builders may be able to co-opt this mechanism to kick-start the inflation of a tiny bubble of space in the lab, making a whole new child cosmos.

Guth obligingly signed the document, and even sat for a photograph holding his statement up with a smile. His friend and colleague, vocal atheist Sean Carroll, then proudly showcased the photo in a public debate on science and God with Christian theologian William Lane Craig. As he recounts the tale to me in his office at MIT, in Cambridge, Massachusetts, Guth seems neither honored nor offended to have had his ideas subjected to such metaphysical scrutiny; he is simply bemused. "I was surprised by the religious emails I got, and continue to get," says Guth.

Now approaching his seventies, with gaunt features and straight gray hair, Guth modestly mocks himself as an "obscure celebrity" whose work is famed in physics circles but whose identity has—for most of his career, at least—lain just outside the mainstream consciousness. I get the impression this quintessential quiet man likes it that way.

This is not the first time we have met. I originally encountered him at a meeting celebrating inflation, run by Stephen Hawking, at Cambridge University back in 2007. Guth is so laid back that he nodded off while sitting in the audience listening to a talk praising his theory. I would later learn that this is a well-known habit.

The religious issues that have come to the foreground later in Guth's life—thrust upon him by others—formed a much less obtrusive backdrop to his childhood. He was born in New Brunswick, New Jersey, where his father owned first a grocery, then a dry cleaning store. His family attended an Orthodox Jewish shul, but only, he explains with a smile, because his uncle was its president. By contrast, his upbringing was as a Reform Jew; his family celebrated the High Holidays but never really observed Shabbat, and he has brought his own children up in a Reform synagogue.

As a boy, Guth was fascinated by the idea of uncovering the hidden rules that govern nature, and loved science. At college he realized physics was the subject that would bring him closest to examining these fundamental laws of nature. While his work now delves into questions that cut to the core of those religious teachings, he says they have played little role in informing his research over the years. "My take on Reform Judaism is that it's a mode of life which accepts the significance of the Jewish heritage and Jewish community," Guth explains. "But I don't read any truly religious ideas into it all."

At the same time, Guth does not dismiss the notion that there may be more in heaven, Earth, and the cosmos than are dreamt of in physicists' philosophies. "Clearly another question that we have good reason to be interested in is, 'Where do the laws of physics come from?'" says Guth. Recognizing this fundamental block in our knowledge is "essential" to his own view on the boundary between religion and science, he tells me. "When we talk about creating a universe in the lab, in this case, or people also write papers about the spontaneous creation of the universe from absolutely nothing as a quantum fluctuation, which is a related question, in all those cases, what physicists do when they approach those problems is to assume that the laws of physics that we already know apply—even when talking about creating the universe from absolutely nothing."

So where might the laws of physics come from? "My current take is we haven't the foggiest idea of where the laws of physics come from. And I think, as a scientist, we have to admit that there is a frontier beyond which we don't know," says Guth. That does not necessarily mean that we need God to fill those gaps, however. "The attitude of most scientists is that the fact that we don't know something yet doesn't mean it is unknowable. We're always trying to push back the frontiers of what science knows, so someday we may understand where the laws of physics come from—as far as I can tell—but I would also say that right now we haven't even a clue about where to look."

I realize that Guth is a master of diplomacy. While he refuses to publicly pick a side, his words are carefully chosen to allow room for both sides in the religion-versus-atheism debate to coexist.

The timing for my visit is peculiar. A few weeks earlier, this mild-mannered physicist, who would rather snooze quietly in the shadows than hog the limelight, was apparently transformed into something like Superman in the public eye when his inflationary model made international headlines. Inflation claims that in its infancy, when the universe was a mere 10^{-34} seconds old, the expansion of the cosmos suddenly sped up, growing at an exponential rate for a short burst, before slowing down again to a gentler pace. It is this same process that, in theory, would breathe fire into the belly of our homemade baby universe, if we ever make one.

That inflationary model is now widely accepted by cosmologists. In Chapter 1 I introduced the cosmic microwave background radiation—the afterglow of the big bang—that pervades the sky, and this has, so far, served as good evidence for inflation. The photons, or particles of light, that make up that radiation have an almost uniform temperature of just 2.7 degrees Celsius above absolute zero, but there are tiny random temperature fluctuations above and below that value in the radiation across the sky. Also as described in Chapter 1, physicists Tony Zee and Stephen Hsu encouraged intrepid astronomers to search for coded structured messages from any potential creator of our universe hidden in a binary code in the arrangement of those slightly colder and slightly hotter spots in the radiation. While no communication from any divine beings or superintelligent aliens who might have made our cosmos has been detected, astronomers have found that the general pattern in the slight temperature differences observed fits the form predicted by inflation theory. Repeated measurements of the radiation with ever more sophisticated satellites—the COBE, WMAP, and Planck probes—have bolstered support for inflation over the years. But the theory also has its detractors, who have offered alternative models to fit the same data.[3]

In March 2014, however, it seemed to the world that inflation had received almost watertight confirmation, finally putting an end to any naysaying. The theory predicts that the violence of the universe's expansion sent ripples through the fabric of spacetime itself, generating waves that would have left a faint but unmistakable imprint on the cosmic microwave background. Any gravitational wave traveling through the infant cosmos would have affected the photons in the radiation that they encountered en route, jostling the orientation in which those photons vibrated in a distinctive way.

It was that signature, the fingerprint of inflation theory, that physicists manning the BICEP2 telescope at the South Pole believed they had found—and they announced the discovery of these so-called primordial gravitational waves to the world at a press conference, with Guth taking a front-row seat as the news was revealed live around the globe. (In Chapter 2, I mentioned the equally vaunted announcement of the discovery of gravitational waves by the LIGO experiment in February 2016. However, there is an important difference between the gravitational waves that LIGO searches for and those BICEP2 is hunting. In LIGO's case, the gravitational waves were generated when two black holes crashed together about 1.3 billion years ago, so that discovery tells us nothing about inflation. Meanwhile, BICEP2 reported the first detection of gravitational waves that had apparently emanated during the inflationary period soon after the big bang. Hence the latter are dubbed "primordial" and can act as evidence for inflation theory.)

The finding was so dramatic that cosmology was all over television news; headlines in the *New York Times* screamed that inflation was correct; and even Guth was excited enough to stay awake through the press conference. Many—though notably not all—cosmologists raved that those touting alternative models to inflation should give up now. Sitting at home, watching this all unfold online, I smiled smugly—the timing of my book could not have been better, I thought, since this groundbreaking result cemented the theory that posits that it may be possible to build a baby universe in the lab.

And then it all turned to dust. Quite literally.

They say a week is a long time in politics. The same, it turns out, could be said for cosmology, which is ironic given the astronomical time scales the discipline usually deals with. Within a matter of days following the announcement, physicists scrutinizing the paper released by the BICEP2 team urged caution, saying that the light signature in the background radiation appearing to show the presence of gravitational waves could have been mimicked by light emitted by galactic dust. It was while the theory was under the shadow of this dust cloud that I actually met with Guth to chat about how he came up with inflation and how it led to the discovery that baby universes perhaps could be created in the lab.

Physically, with his thin, delicate features, glasses, and graying hair, Guth always reminds me of horror writer Stephen King; never having met King, however, I imagine that Guth is less intense. It's possible right now, though, that Guth thinks he is in the middle of his own horror story, having been lauded a matter of weeks ago in the international press but now waiting to see if the BICEP2 result will ultimately crumble. He is amiable about it all, though. By the 1980s, Guth had reached a pessimistic view of the likelihood of being able to engineer a real universe in the lab, despite largely pioneering the idea, and he touched on this pessimism in his popular book *The Inflationary Universe* in 1997. Things have changed now, though, thanks to the work of others we'll meet in the following chapters, and Guth remarks that the possibility of creating a human-made universe has "become more of an open question since the time of the book," adding cheekily, "Like the BICEP2 result!"

It's in this awkward phase, where the BICEP2 results are still tantalizing but not as solid as they appeared to be, that we are meeting. In a few more months, the result would wash away entirely, that particular strand of evidence for inflation lost. I mention this slightly embarrassing failure not in order to deflate inflation, as it were. To my mind—and in the opinion of many neutral cosmologists who have no personal investment in one early universe model versus another—inflation still seems to be the safest bet for describing our early universe. That's not to say that it could never be overturned for a rival model, should one that is better at describing observations come along. But for now—even without the BICEP2 result backing it—inflation still stands as the strongest model yet for the early universe because it solves certain mysteries about the universe and because there is other observational evidence for it.

But building a universe is a speculative idea, hinging on a theory—inflation—that still has question marks hanging over it, and it would be remiss to ignore those. Not least because it was in trying to answer some of these puzzles, most pressingly what triggered inflation to begin, that led Guth and others in the mid-1980s to start pondering whether this process could be triggered again deliberately in the lab, leading to the birth of a whole new universe. That's the story I want to hear about. But first I need to know how Guth—and others independently—stumbled on inflation theory in the first place.[4]

As a cosmologist, I was always taught about inflation a certain way, the same way we are taught most things in physics: there was a mystery (or in inflation's case, three mysteries) about nature that needed to be solved. Someone came along with a model that explained how those perplexing features arose. Then the model predicted something new that had yet to be observed. When that predicted evidence was found by future experiments, the model was accepted. The end.

In inflation's case, that black-and-white, cut-and-dried textbook version of its origin story does it a disservice. Inflation is an odd idea. We saw in Chapter 1 that nothing can move faster than the speed of light (hat tip to Einstein), but inflation says that the universe itself expanded so fast, for a brief period, that space barreled outward at a rate that overtook light. So, for starters, it seems like a crazy speculation. When presented as a solution to certain problems, it just seems too good to be true—like someone sat down and came up with a model that fit the answer in the back of the textbook, without really understanding how or why it happens. It just seems so . . . arbitrary.

The actual story of how Guth came up with inflation, more than three decades ago, is much richer, much less deliberate, and, as a result, far more convincing—since he never set out to solve those mysteries at all. Back in 1979, the young Guth, then a postdoctoral researcher at Cornell, did not even want to be a cosmologist. He specialized in particle physics, trying to understand the building blocks of matter on the tiniest scales, and had only a passing interest in the grander view of cosmology, which concerned itself with tales of how the universe began and where it would end up. He recalls being stunned to learn at a talk that fall by Princeton cosmologist Robert Dicke, one of those who had predicted the form that the cosmic microwave background would take if it was radiation left over from the big bang, that physicists' best ideas of how the universe formed were rife with problems.

Little did Guth realize it at the time, but Dicke was outlining one of the three major puzzles that inflation would soon be invoked to solve: the flatness problem. Simply put, he noted that the expansion rate of the infant universe would have needed to be perfectly tuned—to an accuracy of fourteen decimal places—at one second after the big bang to produce the cosmos we see today. If the universe had been expanding by just one point in the fourteenth decimal place faster, then the universe would have flown apart so fast that galaxies could not have formed. If it had been expanding just one point in the fourteenth decimal place slower, the universe would have recollapsed very quickly and, again, galaxies never would have had a chance to form. But there was nothing in the big bang picture to explain why the universe was constrained to walk this tight rope of "flatness" between the two extremes of either uncontrollable expansion or total collapse.

Now, it could just be a coincidence that the universe hit that precise expansion rate, eventually enabling stars, galaxies, and humans to arise. But physicists tend to hate it when models force them to fix the properties of the universe to values that are happily perfect for life to evolve, something referred to as the fine-tuning of cosmic parameters (an issue that will be explored in more detail in Chapter 7, where we'll meet other examples of how our universe seems to be conveniently tailored for us). Instead, they prefer to seek out mechanisms that would give rise to those particular conditions naturally, without needing a (metaphorical or real) hand of God to make things just right. This discomfort with the idea of fine-tuning can strike a believer as easily as an atheist. After all, even the most devoutly religious physicists study the subject to learn how the universe works, and an answer along the lines of "it's just that way because God made it that way, so don't look any more closely" is deeply unsatisfying.

At this point in the story, we might expect young Guth to be so fascinated by the flatness problem that he dropped everything and came up with inflation to mitigate it. It did indeed seize his attention: "This was all news to me and I found it very shocking," Guth recalls. But, like many other researchers, he then parked the problem in the back of his mind and returned to the—seemingly unrelated—task he had been working on.

Guth was actually busy thinking about an object, predicted to exist by some theories of particle physics, called a magnetic monopole—a kind of a knot in reality with a single magnetic charge. To understand what that means, think about electric charges. An electron has a single negative charge; a proton has a single positive charge. Put them together and they attract, but they can also be separated again. Now think about magnetism. The planet Earth has a north pole (a north magnetic charge) and a south pole (a south magnetic charge), with magnetic field lines running between them. A bar magnet, similarly, has a north pole and a south pole. But if you try to break a bar magnet in half, you won't end up with one piece that has only a north pole and one that has only a south pole; instead you'll have two smaller bar magnets in your hands, each containing a north and south pole. There's no object yet discovered that just contains a single magnetic charge or pole; there are no magnetic monopoles. Why not?

Of course, it could just be that no such thing exists. But in the 1970s—and even today—physicists had good reason to believe, or to at least hope, that they might, and we just haven't spotted them yet. As we saw in Chapter 1, in the nineteenth century physicists had managed to unify electricity and magnetism, showing that they could be described in one theory of electromagnetism. During the latter half of the twentieth century, physicists had hopes that the other fundamental forces of nature could be brought into the same description.

The first to fall into place was the weak force, which governs radioactive decay within atoms. Its role looks starkly different from that of electricity and magnetism. But physicists Sheldon Glashow, Abdus Salam, and Steven Weinberg shared the Nobel Prize in Physics in 1979 for independently realizing that in the furnace of the early universe, these three forces would be combined into one, called the electroweak interaction. As time passed and the universe expanded and cooled, they argued, the cosmos would change phase, akin to water freezing into ice. During that cosmic phase transition, the electroweak force would split into its components, electromagnetism and the weak force. The process would also create new particles, which have since been discovered: the W and the Z bosons.

It's also during phase changes such as these that elementary particles—electrons, quarks, the W and Z bosons, and so on—at first all massless, would gain individual personalities, typified by the fact that they picked up different masses. The mechanism physicists theorized to explain why forces suddenly peel apart and how particles gain mass is based on the idea that the universe is filled with special types of fields named after one of the people who proposed the mechanism, Scottish physicist Peter Higgs. It's these fields that undergo the phase transitions in the universe.

There are different types of Higgs fields, which come with their own special particle, the now famous Higgs boson, which eluded detection for decades.[5] Following the discovery of the Higgs boson at the Large Hadron Collider in 2012, its namesake physicist won a share of the Nobel Prize, along with Belgian physicist François Englert, who also independently developed the theory.[6] The Higgs boson they found, with a mass of about 125 GeV, corresponds to the simplest type of Higgs field that had been predicted to exist.

According to this model, in the very early universe when the temperature is hot, this Higgs field interacts with all particles in a similar manner, and so all particles are massless. While it isn't a perfect analogy, imagine you have a pan of molasses on the stove. When it's hot, it's still thick, but you can drop raisins into it and move your spoon through it relatively easily. Similarly, the hot Higgs field pervading the universe is accommodating to all particles equally, letting them all pass through unimpeded—they have zero mass and just whizz by without resistance. As the temperature cools in your pan, however, the molasses will start to thicken and solidify, and its resistance to whatever ingredients you want to throw in will change. It may also start to discriminate between the utensils you use: a fat wooden spoon will be harder to push through the molasses than a slender metal one. In the universe, there's a much starker and sudden contrast that happens as the cosmos expands, cooling as it does so. When the universe dips below a certain critical temperature, the transition occurs, and the field changes phase sharply. In doing so, the field's interaction with different types of particle shifts. Those that are the most obstructed suddenly acquire more mass, while others, such as the particle of light (the photon), remain massless.

The Higgs boson is what is known as an excitation of the Higgs field. (In a similar manner, in Chapter 2, we saw that photons are said to be excitations of the electromagnetic field.) You can picture this if you imagine the molasses being churned up with more and more energy, perhaps with an electric mixer, creating greater and greater ripples until suddenly a glob of molasses jumps out of the surface, then quickly drops back into the mixture to recombine with the rest. The Higgs particle evaded detection for so long because it takes so much energy to excite one out of the Higgs field.

After the success of unifying the electroweak force, the hunt was on for a grand unified theory that would bring into the fold the strong nuclear force, which binds quarks together in atomic nuclei. According to these speculative models, the electroweak and strong nuclear forces were combined into one über-force in the infant universe. Grand unified theories also employ Higgs fields with their own Higgs particles—hypothetical cousins of the one detected in 2012. When one of these fields changes phase, it is proposed, an initially unified single force splits, eventually branching into the separate electromagnetic, weak, and strong forces we see today.

This is the long way of getting us to monopoles. Such phase changes in the fields that stretch across the universe would be so cataclysmic that they would leave potentially detectable scars on the universe—some of which would be monopoles. More familiar phase transitions leave similar scars, or defects. Ice in your freezer is usually cloudy rather than perfectly transparent because when water solidifies under certain conditions it contains defects, or regions where the water molecules are misaligned. In the same way, physicists proposed that there could be cosmic defects scattered throughout space, left behind because portions of the Higgs field set in a haphazard way after the phase transition. One possible type of defect would be a magnetic monopole, existing like a bubble in an ice cube. But whereas the bubble in the ice is filled with something—tangible air—the monopole is a spherical knot in space itself, with magnetic field lines pointing out symmetrically from it, like the husk of a chestnut or a curled-up hedgehog.

Monopoles are predicted to exist by grand unified theories. Yet no monopole has ever been detected. What's worse is that some models suggest that if they do exist, they would be so heavy that their gravitational pull would draw all matter together again so strongly that the universe would collapse inward. This conundrum of the missing monopoles was on the minds of many particle physicists in 1979, when Guth was a postdoctoral researcher at Cornell. If the next level of the unifying theory was correct, they should exist, but where were they? And why had they not destroyed the universe?

It wouldn't be fair to say that these questions fascinated Guth; rather, it was a friend and fellow postdoc, Henry Tye, who seized hold of these puzzles and did not want to let go. Tye came to Guth to see if he had any insight on the relationship between monopoles and grand unified theories. Guth had thought so little about the issue that Tye ended up having to explain to him what grand unified theories actually were. But with that information, Guth knew enough about monopoles to carry out a calculation confirming that if these theories were correct, magnetic monopoles should exist—but they should also be "ridiculously heavy," as he puts it. Particle physicists usually try to generate new predicted particles in accelerators, huge machines that smash everyday particles together at high energies in the hope that they will break apart into their constituent pieces and exotic new particles will form in the debris. But Guth's quick calculations showed that magnetic monopoles would be so heavy that there was little hope that accelerators would ever be able to generate the energies needed to produce them. "I told Henry he should just forget about it," says Guth.

"There's no way we'll ever see these magnetic monopoles, or be able to tell if they exist or not."

Luckily for Guth, Tye was a tough man to dissuade. He remained intrigued by grand unified theories and felt that their connection with monopoles was a new way to attack the question of whether they could be right. Discovering a monopole would provide evidence for the still speculative grand unified models—quite a coup. On the other hand, if Tye and Guth could show conclusively through their calculations that monopoles simply could not exist, it would rule out grand unified theories. So Tye challenged Guth to help him work out just how many monopoles should exist if grand unified theories hold true.

Guth was resistant. "It sounded a bit crazy to me," he admits. Grand unified theories were speculative to start off with, and Tye was asking him to look into the consequences for the early universe of such models being true. This was the realm of cosmology—an area Guth knew little about. Back in the 1970s, cosmology was hardly a precision science, with no large-scale experiments or observations to test suppositions of what the infant cosmos might have been like. Bringing grand unified theories together with cosmology was like speculation squared. "I didn't think there was very much to be learned by combining these two very uncertain topics. It was clear early on that if we were going to be talking about the production of monopoles in the early universe, in the context of grand unified theories, I would be talking about ridiculous time scales, like 10^{-35} seconds after the beginning, which at that time sounded totally crazy to me," says Guth. He chuckles, acknowledging that the work he is now famed for today sits squarely within that minuscule time frame.

For the next six months . . . well, nothing happened. Tye was working on other stuff, but he continued to push Guth to start to think about the question, and Guth continued to push back. But then something changed his mind. Weinberg, the behemoth of particle physics who had been one of the physicists to unify electromagnetism and the weak force, visited Cornell to give a series of lectures about applying grand unified theories to cosmology, extrapolating the history of the universe back to that dreaded time window of just 10^{-35} seconds after the big bang—essentially what Tye had been failing to persuade Guth to do for the past half year.

This was a matter of months before Weinberg was honored by the Nobel committee, and Guth suddenly realized that if such a towering figure thought there was serious insight to be gained by combining these two areas of physics, then the young Guth was certainly not qualified to scoff at his decision. "Really, immediately after Steve's visit, Henry and I started working on the magnetic monopole problem seriously," says Guth.

Once they were engaged in the project, it did not take the pair long to come up with a pretty staggering result: if grand unified theories were correct, the universe should be "swimming with magnetic monopoles"—with roughly one monopole for every proton that currently exists. Since Guth had calculated that monopoles must also weigh in at a whopping 10^{16} times the mass of the proton, that would fantastically change the estimated density of matter in the universe.

This led to other ridiculous predictions. The density of matter in the universe is intimately tied to calculations of its age. That's because after its explosive birth, the universe expanded outward, but that growth was slowed down by all the mass within the universe, whose gravity acts to try to pull all the mass back together again. Cosmologists can estimate the age of the universe by observing how fast it is expanding today and working out how quickly the matter in the universe could have slowed down that rate after the big bang. If the density of matter in the universe is dramatically bigger—as it surely would have to be if monopoles existed in the quantity suggested by Guth and Tye's calculations—then the universe would have slowed to its present expansion rate far sooner than had been thought previously. When they did the math, their calculated age of the universe came out to be just 10,000 years, "which is nearly biblical," Guth laughs, "but not very geological, so we did not consider that to be a possibility."

The finding seemed so absurd—and rather disappointing for fans of grand unified theories—that the pair did not rush to publish it. That was a mistake, because they were scooped on publication by physicist John Preskill, now at Caltech, who was then a graduate student working with Weinberg at Harvard. Guth and Tye were devastated, and they scolded themselves for not working faster, as they surely would have if they had known that Preskill was on their tail. "We were panicked and aggravated and disappointed," recalls Guth.

There was more at stake than just the frustration of not being able to publish one paper on monopoles. At this time, Guth had begun a family with his wife, Susan, and their first son was not quite a year old. But Guth was struggling to find a steady university job. Like so many young graduates, he had become stuck in limbo, moving through a series of postdoctoral research positions. These temporary positions usually last for a couple of years and serve as apprenticeships for newly minted PhDs in science to learn research skills before settling into a faculty position. In theory, there is much to be gained by such a system: it allows young academics the opportunity to travel around the world, honing their skills with various mentors, and gives them the freedom to turn their hand to anything that interests them, without the shackles of a PhD supervisor dictating their research direction or the teaching and administrative responsibilities that distract university professors. But the downside is that these researchers, who are often starting families, do not have secure jobs, are working for relatively little money, and cannot settle in one location, knowing that they will likely be uprooted to a new job—possibly in a different country—in a matter of months.

Plus there is no guarantee that after paying your dues as a postdoc you will have published enough work of a high enough standard to win yourself a faculty position, should an opening even appear. It is a precarious life, and a large reason why I chose to leave academia after completing my PhD; I did not want to go through the years of insecurity. Guth's situation by this time was pretty grim, as he had been a postdoctoral researcher for eight years—"which is a very long time, practically a record," Guth chuckles. He had succeeded in getting temporary placements at Princeton, Columbia, and Cornell, so his CV certainly included all the right names. "But still I didn't have a faculty job, so it wasn't clear what the future was going to be," he says. Guth needed to prove his mettle soon and earn himself a faculty position or else leave academia, and he could not afford to waste months of work on research that could not be published because he had been beaten by a competitor.

So, having put in so much effort, Tye and Guth scratched their heads to see if they could salvage anything. It was not worth trying to simply publish their work now, which appeared to only replicate Preskill's. For it to be worth something, they had to come up with something more, something new. Both they and Preskill had shown that grand unified theories, as they stood, were not compatible with cosmology. The next natural question to ask was whether grand unified theories could be slightly modified to get rid of the pesky monopole problem and make them fit with our universe's history. Though he did not know it at the time, this question would soon turn Guth's life upside down, because it would lead him and Tye to inadvertently formulate what he calls the "entree to inflation."

In an effort to artificially prolong the amount of time that he was employed at Cornell, Guth requested a year's leave of absence to visit the Stanford Linear Accelerator (SLAC), in California, the site where various elementary particles had been discovered over the years. While there, he continued to work with Tye, trying to think of mechanisms that might have suppressed monopole production in the early universe. If they could come up with one, it might explain how grand unified theories could still hold without the cosmos being plagued by massive monopoles supposedly lurking around every corner. But now that they were geographically separated, working with Tye became harder. There was as yet no Internet and no email, and so they updated each other on their progress over the phone.

This was the problem as they saw it: Grand unified theories predict that when the universe was younger, hotter, and denser, the electromagnetic, weak, and strong forces were unified into one grand force. The theory says that the universe is pervaded by Higgs fields and that as the temperature of the cosmos cools, the fields change phase. At these cataclysmic phase transitions, the forces separate and new particles are created. The transition leaves scars on the universe, including the knots in the Higgs fields called magnetic monopoles, which form at the boundaries between regions where the Higgs field had misaligned. The trouble with this picture was that the phase transitions that produced monopoles happened at a relatively high critical temperature, so even though the universe was in the process of expanding and cooling down, it was still pretty hot. This meant that it was full of energy and extremely chaotic, making it more likely to produce knots. Think of the universe like an old-fashioned telephone cord or a rope: the more you jiggle it, the more likely it is to get tangled and knotted. So in this case, far too many magnetic monopoles will be formed, in theory—a haul that simply is not seen in the universe today. Since this crucial monopole prediction is not borne out, it seemed at first glance that grand unified theories must be wrong.

But what if there was some way to delay the monopole-producing phase transition to a time when the universe was cooler? Then, like a telephone cord that is only rarely and gently jostled, the Higgs fields would have far lower energy and would be less prone to knotting. There is a mechanism that can delay the temperature of phase transitions, an example of which most of us see every day when we step outside and look into the sky: supercooling. Clouds at high altitudes are suspensions of water droplets that exist as supercooled liquid at temperatures well below water's freezing point. Water droplets usually crystallize to ice at 0 degrees Celsius (32 degrees Fahrenheit) because they have a piece of grit to act as a seed around which to solidify. But in clouds, without that grit, water in its purest form can resist freezing down to -40 degrees C (-40 degrees F). Guth and Tye realized that something similar could be true for the Higgs fields, postponing their transition to a cooler temperature and reducing the number of monopoles formed to a more manageable number.

It seemed so simple in hindsight. Physicists knew about both phase transitions and supercooling already—and they understood that phase transitions would have occurred in the early universe—but somehow nobody had thought of bringing the idea of supercooling into cosmology. The problem had seemingly been solved, apart from one loose end that popped into Tye's mind: would the supercooling have any noticeable effect on the expansion rate of the universe?

Guth had assumed the supercooling would have no effect, but at Tye's urging, he went home to check further. He remembers that evening well because the discoveries rolled in thick and fast. It was December 6, 1979, and by the end of the night, Guth had discovered inflation and solved one of the biggest problems in cosmology.

To understand why, we have to know a little more about how energy can be stored in the Higgs field—or indeed in any field. The simplest example of a field is a weather map marking different temperatures across a large area. Just as a sun seeker could use a global weather map to plot out a round-the-world voyage that would stop at all the hottest destinations, Guth needed his own field map that would take an adventurous universe on a journey between states of different energies. In particular, Guth needed to plot a journey in which the universe would have a late phase transition thanks to supercooling—reducing the number of monopoles produced in the cosmos, as he and Tye required—and then use the plot to see what energy the universe would have had before, during, and after the transition. Knowing these energies would help him to calculate the rate at which the cosmos expanded during and after the transition.

Particle physicists call the lowest energy state of a field the vacuum. As we saw in Chapter 2, quantum uncertainty can allow odd things to happen. Just as you can never quite pinpoint the position of a particle before you look at it, you cannot even really be sure that there's nothing there. Even in a seemingly empty field there is some probability that two virtual particles can pop up, as long as their properties are equal and opposite. For instance, a negatively charged electron could appear instantaneously along with its corresponding antiparticle, the positron, which has an identical mass but a positive charge. The particle and antiparticle pair are living on borrowed time, however, and can only stay for a fleeting moment before they combine, annihilating each other and vanishing into nothingness again.

Despite this constant roiling of virtual particles, overall the energy of the vacuum is bubbling around zero.[7] If the universe settles into this vacuum state, then it's in a pretty stable position. We know that low energies are more stable—it's why we are more likely to build a house in the valley between two hills rather than halfway up the hill, where there's a risk that your house will go sliding down to a point that's lower in the gravitational field.

But there's another option for house building: you could set up your home at the top of the hill if there is a plateau at the summit. You could even picture a situation where at the top of the hill there's a big dimple. If you lived in a house nestled in that small dip, you might even be fooled into thinking you lived in the stable valley, because when you looked out of the window you would see the slope of the ground rising up in all directions around you. That dimple would be a fake minimum, where you could sit comfortably without worrying about rolling down the hill.

It turned out that the Higgs fields that interested Guth and Tye had an energy profile that was similar to a dimpled hill. The lowest energy state for the Higgs field—its true vacuum state—lay at the bottom of the hill, in the circle surrounding the hill's base. In most cases, if the Higgs field ended up in an excited state above this energy, it would land somewhere on the slope of that hill, making it prone to roll down to the bottom again, minimizing its energy. But there was also a chance that the Higgs field could end up with a high enough energy that it settled into the dimple at the top of the hill. Although it was still excited into a high energy state, it was relatively stable because it would take a big energy kick to scale the walls of the dimple, getting it out of that mini-rut. This high-energy state within the dimple is called a false vacuum because even though it isn't actually the lowest energy state—that's the true vacuum—it is a relatively stable energy state.

Under normal conditions, without any supercooling, the grand unified theory phase transition corresponds to the Higgs field moving from a high-energy state to stability at the bottom of the hill—the true vacuum. The temperature of the phase transition is high, which causes the energy of the Higgs fields to jump around violently. Even if the field gets momentarily stuck in the dimple, there's enough energy to cause it to jump out of the dimple and then roll down the hill to the true vacuum. But what Guth quickly realized that night when he carried out his calculations was that if there was supercooling in the early universe, as he and Tye hoped, then this prediction would change drastically. If the temperature dropped dramatically before the phase transition happened, it was likely that the Higgs field would get trapped in the false vacuum for much longer, because the background temperature would not give it enough energy to hurdle over the wall surrounding the dimple and make it to the true vacuum at the bottom of the hill. So even after the temperature fell well below the normal temperature of the phase transition, the Higgs field—the universe itself—would be stuck in this high-energy false-vacuum state.

This is where things get interesting for the infant universe, because false vacuums have some odd properties. Remember that if the universe was in the true-vacuum state, its energy density—the amount of energy in one cubic meter of space, say—would be more or less zero. But in the false-vacuum state, the energy density is quite high. More important, that energy density is fixed, so if you stretch out your cubic meter of space to a volume that's twice its size, the new, larger volume would have the same energy density.

That in itself may not sound too dramatic, though it is completely different from what we are used to seeing in everyday life—bringing with it some very unusual consequences. For example, we could fill a small balloon with helium gas. If we then pulled on the walls of the balloon to double its size, the density of helium inside the balloon would be halved, because the same number of particles now occupies double the volume. And since the energy of the helium atoms inside the balloon has not changed, the energy density in the balloon would be halved too.

Now think about what happens if we fill an imaginary balloon with false vacuum instead of helium and then once again pull out the walls of the balloon to increase its volume to twice the original size. Since the Higgs field always has the same energy density but the volume has doubled, then the amount of energy inside the balloon must also have doubled to compensate. Here comes the kicker: In Chapter 1, we saw that Einstein's general theory of relativity showed that gravity manifests because spacetime warps around heavy objects. (In our solar system, the dominant mass is the Sun, which bends spacetime around it and holds the planets in orbit.)

But energy can curve spacetime and create a gravitational field too. In the case of our false vacuum, this replenishing energy input creates an outward gravitational push, forcing the balloon's walls outward. Now, turning back to the cosmic consequences of a false vacuum in our already expanding infant universe, Guth realized the false vacuum would have driven the cosmos to expand even faster.

And Guth's calculations showed that the false vacuum would not just speed up expansion a little bit but instead would inflate the universe at an exponential rate. In just 10^{-37} seconds, the size of the universe would double. Wait another 10^{-37} seconds and it would be four times the size. In the next 10^{-37} seconds it would be eight times the size, and so on. Within just 10^{-35} seconds—an unimaginably tiny amount of time—space would have doubled in size a hundred times over, becoming 2^{100} times larger (1,267,650, 600,228,229,401,496,703,205,376 times bigger). For comparison, without this rapid inflation, space would have grown just ten times larger in that time. "It is actually pretty obvious once you write down the equations—it's not very subtle—that if the universe had undergone this supercooling . . . that would drive exponential expansion," says Guth. "That would radically change the expansion history of the universe."

I stop Guth midflow. His calculations show that space would be inflating so rapidly that the expansion would outrun the speed of light. Didn't that give him pause? Einstein's special theory of relativity, which we met in Chapter 1, tells us that no object can ever catch up with light. Was he worried that what he was predicting for the universe simply could not happen?

Guth, however, was not perturbed, noting that Einstein's rules apply only to the motion of matter in space, not to space itself. Rather than contradicting Einstein, says Guth, "inflation is something which very much relies on the real guts of general relativity. It makes use of the idea that space is plastic." General relativity allows for the fact that the space between two stationary objects can stretch over time, flinging those objects apart at faster-than-light speed without those objects actually needing to move relative to space at all.

Inflation would not go on indefinitely, however. If only classical rules of physics applied, then a universe settled into the dimple at the top of the energy hill may stay in its false-vacuum state forever, because it would never have enough energy to climb over the walls of its rut. But quantum rules, as we saw in Chapter 2, have a bit more wiggle room because there is always some small but significant probability that the universe will just tunnel out of the dimple and find itself rolling down the hill into the true vacuum at the bottom. At this point inflation stops, having lasted for no more than a fraction of a fraction of a second, but having had a dramatic effect on cosmic history.

That same night, Guth's mind turned back to the lecture by Dicke he had heard a year earlier, talking about the flatness problem. This was the puzzling fact that the rate of expansion of the early cosmos apparently needed to be precisely tuned to just the right value to create a universe with the characteristics needed to evolve galaxies and life. The expansion rate was tied to the density of matter in the early universe; the denser the matter, the greater the inward pull of gravity countering cosmic expansion. This, in turn, controlled the universe's geometry to a fine degree. A tiny sway in the numbers one way, giving a slightly less dense infant cosmos, would likely generate a negatively curved or saddle-shaped universe, in which the cosmic expansion hurtled along so fast that it would rip clumps of matter apart before they could settle to form galaxies. If the cosmos started out slightly too dense, however, the extra inward gravitational tug would cause the geometry of spacetime to curve inward on itself, rapidly collapsing the cosmos back down and again preventing galaxies from forming.

Dicke had noted that cosmologists had already identified the critical matter density needed in the early universe that marked the knife edge between these two extreme geometries—it designated a spacetime that did not curve either inward or outward but was quite literally flat. Our early universe seemed to have been tuned to have this critical matter density, the perfect conditions for evolving life, but it was perplexing as to how this would have happened. Guth suddenly realized he had inadvertently stumbled on a mechanism—inflation—that could solve the puzzle of how our universe ended up with a flat geometry without needing to fine-tune the cosmos's initial matter density in a contrived manner.

When you factored in inflation, the universe did not have to start out at critical density—in fact, it could even be very far from it, either much greater or much smaller. And it could have any geometry you liked; prior to the onset of inflation, the cosmos could have a saddle-shaped geometry, or one that was closing in on itself. The rapid inflation in size of the universe had the effect of automatically flattening out the geometry of any small patch of this cosmos. The mechanism for doing so was no more complicated than if you look at the curvature of a small section of a balloon as you blow it up; the more the balloon expands, the flatter any small region on its surface appears to become. "The exponential expansion drives the universe toward flatness," says Guth.

All of this hit Guth rapidly that night. I ask him what he did next: rush to tell his wife, dance around the room with joy, cheer? . . . "All those, I think," he tells me with a laugh. "I was very excited. I did tell my wife. I don't know how well that came across—I don't think very well, actually." But then doubts began to creep in. "I certainly thought I had made a mistake. I was very leery of it at the beginning," he recalls. "I thought it was too good to be true—and things that are too good to be true usually are not. That's always been my experience."

The next morning, as Guth raced to his office at SLAC to double-check his calculations, he broke his personal bike speed record, getting to his office in a time of 9 minutes and 32 seconds, which he used to note down in his diary each day. At his desk he took out a sheet of paper and wrote down: "SPECTACULAR REALIZATION: This kind of supercooling can explain why the universe today is so incredibly flat—and therefore resolve the fine-tuning paradox pointed out by Bob Dicke." For added emphasis he drew a double box around those words, and then went back over his math. The numbers seemed to check out.

Frantically he called up Tye, ready to share his eureka moment. But, says Guth, "I was disappointed when I called Henry because I was unable to get him excited about it." Tye's lack of enthusiasm nagged at Guth in the intervening decades. Funnily enough, he only learned the reason for Tye's muted response a few days before our meeting, thanks to an interview Tye gave to the *Boston Globe* about inflation, in response to the hype around the BICEP2 results. In the interview, Tye explains that back then he did not actually understand the flatness problem and thus did not appreciate the importance of finding its solution. Guth shakes with genuine laughter as he tells me this. "Which makes everything make sense, because it really did not make sense to me previously," he says. "It makes me feel less bad!"

Tye's lack of strong interest coupled with the fact that he was about to head off to China for a six-week break left Tye reluctant to pursue inflation further. Before leaving, though, Tye was keen to push through the definite result that they had—that supercooling could dispel the monopole problem—without worrying about the consequences of an inflationary effect on the universe. Tye's sense of urgency was compounded by the fact that Guth had developed a reluctance to submit papers prematurely that bordered on neurosis. It's a habit that is still with him today and exasperates his co-authors, since he freely admits he tends to delay publication, sometimes for years, as he painstakingly checks and rechecks his work. "My collaborators still get very frustrated about it," Guth admits, sounding slightly embarrassed. "I just don't feel like sending out a paper until I am sure that every equation is right and every *i* is dotted and every sentence is correctly punctuated."

So Tye had good reason to want to drop the work about inflation and concentrate purely on the supercooling, which the pair did, pushing a paper through for submission to the journal *Physical Review Letters*. Then, as Tye was leaving for China, Guth asked Tye's permission to explore inflation on his own, perhaps even writing his own paper about it. The six weeks until Tye returned and they could jointly pick up this promising topic seemed like an "eternity" to Guth, especially since there was really no way the pair could collaborate remotely between North America and the Far East at the time. Tye agreed, dropping any claim on inflation just as Guth was about to discover that his new theory solved another mysterious cosmic conundrum.

I stop Guth here and ask him if, with hindsight, Tye has looked back on that decision ruefully. "I think he regrets it," Guth replies quietly, but adds, "He's very gentlemanly about it. He doesn't complain to me about it at all."

Now working alone, Guth ran with the idea, helped again by another fortuitous encounter with cosmologists. A few weeks after Tye had departed, Guth happened into the SLAC cafeteria for lunch and ended up caught up in a conversation about work by Zee. In a new paper, Zee had proposed some solutions to what was known as the horizon problem, another cosmological puzzle that Guth knew nothing about, so he asked his friends to explain.

The horizon problem arises because the entire sky appears to be pretty similar, no matter where we look—but cosmologists cannot explain why. In particular, the cosmic microwave background has almost exactly the same temperature everywhere. After the violence of the big bang, it would not be likely that radiation would be so perfectly uniform across different regions—unless those regions had been in close contact for long enough that their temperatures would even out, like the coffee in your cup cooling to room temperature if you leave it out and forget to drink it.

But here's the problem. Radiation can even out its temperature between two points only if information about the temperature has had enough time to pass between those two points, conveyed by photons in the radiation zip-ping back and forth. That time is limited by light's finite speed, which also constrains how far and fast information can travel. Thanks to light's speed limit, when you look out into the night sky, you cannot see infinitely into space. Even with an imaginary perfect telescope, if you look out at light coming from the west, you will be able to see only as far as light has had time to travel since the big bang, almost 14 billion years ago, which fixes a horizon beyond which you cannot see. There may well be much more uni-verse beyond the horizon, but for now, it is shrouded from us. As each day passes, a little more of the veil is lifted and the horizon spreads out further.

Now, roughly speaking, if you pick out a galaxy 10 billion light-years to the west and another galaxy 10 billion light-years to the east, those two galaxies are separated from each other by 20 billion light-years—meaning that it would take light (and temperature information) at least 20 billion years to pass between them and even out the differences in temperature between the galaxies. But the universe has existed for only 13.8 billion years, which should not allow enough time for any information to have passed between those galaxies.

This could be seen even more starkly when Guth crunched the numbers involved with the cosmic microwave background. Astronomers had seen light points at opposite ends of today's sky that were emitted when the universe was 380,000 years old (remember from Chapter 1 that the cosmic microwave background serves as a "selfie" that the universe took of itself around 380,000 years after the big bang). Furthermore, they can calcu-late that the light we are seeing from those two sources was transmitted from two spots lying as far as 90 million light-years apart, and that those spots have almost the same temperature. But back when the universe was 380,000 years old, the distance from one horizon to the opposite horizon would have been just 900,000 light-years. (You might think that it would be 600,000 light-years—300,000 from the origin point in one direction and 300,000 in the other—but it's greater because space was also known to be expanding, carrying light further along with it.) So somehow the tempera-ture between those two spots evened out over a distance that is a hundred times greater than the distance light could have traveled in that time.

Guth quickly realized that inflation provided a pretty simple expla-nation for the horizon problem too. Inflation's whole role was to take a tiny patch of the universe soon after the big bang—a patch that was small enough to have easily reached a uniform temperature from end to end—and blow it up exponentially, faster than the speed of light, so that it stretches over a far larger region than would be expected from the big bang model alone. That meant that astronomers looking at the furthest reaches of the sky would be looking at light from distant points that had been in close contact at the beginning of the universe—hence having the same temperature.

If that wasn't enough, inflation had one more trick up its sleeve: it seemed to neatly solve once and for all the mystery that Guth had started with, which was why nobody has ever found a monopole. Guth and Tye's initial analysis had already shown that the supercooling that led to inflation would automatically reduce the number of monopoles you might expect to find in the universe. This was good news: Guth's previous calculations had shown that monopoles were pretty hefty, and so if there were a huge number of them hanging around, they would gravitationally tug everything else in the cosmos inward, destroying the universe. But even given this reduction in monopole numbers, there still should be *some* monopoles, and the fact remained that none had been seen. Inflation provided the answer: there may well have been a reasonable number of monopoles dotted around space soon after the big bang, but at such a low density that you could imagine finding a small patch of space with just one monopole in it. If inflation took that patch and blew it up into the uniform cosmos we see around us, then the monopole density within our horizon would have become so diluted it was possible that there may be just one monopole lying within the visible universe. If so, then the odds of us stumbling on it are next to zero.

Guth was thrilled. In one stroke inflation had solved three major cosmic conundrums: the flatness problem, the horizon problem, and the monopole problem. In January 1980 he went public with the idea, giving a talk at SLAC. Word began to spread, first in the particle physics community and later through the astrophysics community, and Guth found himself on a lecture circuit. He'd spent almost a decade struggling to get a faculty position, and now suddenly everyone seemed to be offering him a job. Well, everyone except his first choice, MIT, a place that he felt was his natural home, since he'd been to grad school there.

The quiet man decided that the MIT job was the one he wanted, but his natural shyness was holding him back. Then, on the last day of a trip visiting colleges around the country, while at the University of Maryland, he went to a Chinese restaurant and got a fortune inside a cookie that he has kept ever since: it said, "An exciting opportunity awaits you if you're not too timid." Sounding almost like a naughty schoolboy who has just decided on a prank to play on a teacher, he recalls that he decided at that moment to call up one of MIT's senior physicists, Jeffrey Goldstone. With his trademark unassuming politeness, Guth told Goldstone, "Well, as you know, I have a number of different job offers, but if MIT made me a job offer, that would probably be the one I'd be most interested in accepting." He chuckles sheepishly as he recounts this. The next day, Arthur Kerman, then the director of MIT's Center for Theoretical Physics, called with a job offer. Guth was ecstatic.

I ask how far that newfound confidence has carried through into the rest of his life since then. "I've certainly been a little quicker to ask for what I want than I was in the old days," he says with a laugh. But, he adds, "I guess I still consider myself a shy person."

So by early 1980, inflation had rewritten the universe's past, and Guth's future. He finally had a place to settle with his wife and young child. But despite his excitement, there was something niggling in the back of his mind about inflation. "I was still worried that there was something catastrophically wrong about it that I hadn't realized yet," says Guth.

It turns out there was. Inflation theory, as Guth had outlined it, could indeed solve a host of problems with the big bang model of universe, but at a cost. Inflation was meant to last for just a fraction of a moment after the big bang before the inflating false vacuum turned into the stable true vacuum of the universe today. However, when Guth looked more closely at his model to see how this would happen, he realized that, according to his model, inflation would never stop.

The solution to that problem would come from a young physicist living on the other side of the Iron Curtain—a man who, bizarrely, had been working on a theory of telepathy and thinking about the role of a supernatural consciousness in the universe before independently discovering "new inflation," and later how inflation may lead to a multiverse of many parallel universes and a possible way to make a universe in the laboratory: Andrei Linde.

4

Bursting Inflation's Bubble

Andrei Linde is a showman: bombastic, passionate, and fueled by the certain belief that inflation theory, which he helped to invent, is correct. If you want to find Linde at a conference talk on inflation, wait until the speaker has finished; while Alan Guth, whom we met in Chapter 3, is often to be found snoozing, Linde will almost invariably stand up and comment—loudly. Woe betide speakers who dare to criticize inflation. Linde, a tall, broad-shouldered man in his late sixties with a full head of silver hair, cuts an imposing physical presence, and will dress down critics in no uncertain terms in his thick Russian accent.

This obstinate refusal to let any criticism pass unchecked has afforded him an intimidating image. Many of my contemporaries were too scared to approach or speak to him when they were young postdoctoral researchers. But there is also a playful and humorous side to Linde, and he will often entertain conference attendees with his magic tricks, complete with banter. (One of his favorites is showing off his "psychic" ability to guess which hand a volunteer raised while his back is turned.) Among his colleagues he is well liked, and his commitment to following his hunches—even if they seem to be leading him in an absurd direction—has led him to not only revolutionize our understanding of the infant cosmos but also wrestle with some deep philosophical issues that many physicists shy away from.

I am meeting Linde at his office at Stanford University, in California, to talk about his work on inflation and—eventually—baby universe creation. It's Linde who wrote the paper that partially inspired Tony Zee and

Steve Hsu, whom we met in Chapter 1, to start pondering about ways to look for signs of a cosmic creator, spurred by Linde's suggestion that our own cosmos was made by a "physicist hacker who baked the universe in his own laboratory." When Linde submitted his draft, titled "Hard Art of the Universe Creation," to the journal *Nuclear Physics B*, in 1991, he received an angry referee report saying that the physics was fine, but the title of the paper, its abstract, and "this dirty joke [at the end of the paper] about the universe being made by a physicist hacker" must all be changed or removed because they "may offend religiously minded people." Linde partially acquiesced, giving the paper the more sober title "Stochastic Approach to Tunneling and Baby Universe Creation" and a dryer, more technical abstract that was less likely to stir emotions—or, indeed, as much interest.[1] I think this addresses one of the questions I posed in Chapter 1: how is it that the potential ramifications of the proposed project—to make a new baby universe in the lab—largely passed physicists by for so many years? The answer is that many physicists either self-censored the implications, as Zee did, or were politely asked to tone down their claims, as Linde was. But Linde stuck to his guns about his reference to the physicist hacker, writing back: "I am not going to remove the joke from the end because I am not so sure that this is just a joke."

I am jumping way ahead here, though. Before Linde and I can get to the subject of lab-made universes, I need to know about his profound doubts about the nature of reality that drove him to study cosmology. Eventually this probing line of thought led him to solve the problem that Guth came up against in Chapter 3 and to develop the theory of "new inflation."[2]

Linde, the child of two physicists, was born and raised in Moscow when Russia was still part of the Soviet Union. His mother, now in her nineties, worked at Moscow State University, studying cosmic rays, and was also commander and chief of staff of a female division of night bombers during the Second World War, while Linde's father was a radio physicist. With two parents in physics, Linde's fate might have seemed set. In fact, by the time he was ten, his ambition was to become a geologist, and he carried around a rock hammer and filled his backpack with stones to build his strength for future rock-finding expeditions. But then, just after he graduated from seventh grade, his family went on vacation to the Black Sea. For the week-long car journey, they gave him a book about astrophysics and another popular book on special relativity. "When we finally arrived at the Black Sea, I felt terrible," says Linde. The child had prided himself on his commitment to geology. "I was not very good at volleyball or other sports—so what was good about me? [Geology] was my future," says Linde. "Everybody in my class knew that I was a geologist, and now I was going to come to them and say, 'Sorry, guys, I was lying to you—I changed my mind!'"

But change his mind he had. From there on he focused on physics, with some time taken out for reading science fiction with his gang of friends at school and pondering philosophy. With his book club, he read a *Matrix*-style story by Polish author Stanislaw Lem, about people discovering they were actually brains trapped in vats and that external reality was an illusion. Linde found himself fascinated by deep questions of what really exists and how we could know it.

Linde's leanings were out of kilter with the prevailing Soviet attitude, however. In communist Russia, religion was largely frowned upon, and Linde had not been brought up in any faith. He still considers himself outside of such strictures, but jokes that, "with my strange philosophy, I was the most religious person around." He rejected many of the secular values his fellow citizens were supposed to espouse. "Vulgar materialism" still irritates Linde, who read Lenin's famous book *Materialism and Empirio-Criticism* as a young adult but quickly dismissed it as "very witty, but not very wise."

Throughout his career Linde has remained fascinated by the role of consciousness and—perhaps surprisingly—what it may mean for the origin of the universe, arguing that science can never claim to be complete until we can explain the concepts of life and the mind. "It is possible to postpone it, it is possible to ignore it, it is possible to reduce it to cybernetics, to chemical reactions and electromagnetic reactions in the brain," he says firmly. "But science is absolutely, utterly incomplete until we understand at a deep level what we are talking about with the reality of consciousness." Without that understanding, he adds, "we will never understand the reality of the universe."

To explain how consciousness might be connected to the origin of the universe, Linde talks me through a series of lectures on the philosophy of cosmology that he recently gave at Hamburg University in Germany. In them, he drew parallels between quantum cosmology and Indian philosophy, and to me he admits that he was "scared" to go public with his thoughts about connecting aspects of physics and spirituality.

Quantum cosmology is an attempt to bring quantum theory, the physics of the very small, together with general relativity, which holds sway over the very large, to provide an ultimate description of the universe. In Chapter 2 we saw that it's impossible to predict the outcome of quantum experiments with absolute certainty. But in the 1920s, Austrian physicist Erwin Schrödinger had come up with an equation to calculate the probable outcomes of any quantum experiment. Within Schrödinger's equation, quantum systems are characterized by their wavefunctions—a mathematical term that encapsulates this probabilistic nature of the quantum world. In the 1960s, American physicist Bryce DeWitt, influenced by Zee's one-time mentor John Wheeler, audaciously attempted to come up with a master wavefunction that governed the whole universe. At first blush that may seem like an odd thing to do, since quantum laws are usually thought to have an important influence only on minute scales, which is where wavefunctions typically are assumed to operate, and by almost anyone's reckoning the entire universe today is anything but small. But the universe started out tiny, early in its cosmological life, and there quantum physics cannot be ignored. So if physicists want to understand how the universe as a whole behaved in the moments after the big bang, they must consider quantum effects on the cosmos. And since all of the universe's matter is squeezed into this small infant cosmos, the influence of gravity is crucial too. Rather than just using Schrödinger's equation alone, then, DeWitt had to combine it with Einstein's equations of general relativity.

But when he did so, Linde explains, an odd thing happened. The math of both general relativity and quantum theory feature time as a variable. They must do so, or physicists would not be able to use them to predict how things should change as time passes. But the Wheeler-DeWitt equation, which married the quantum and the cosmic, showed that the rate at which the universe evolves is zero; in other words, the universe is static. In Chapter 1, we saw that Einstein favored a stationary universe that neither expanded nor contracted, but this went much further than that. Time had completely vanished from the universe, suggesting that everything within it should be frozen too. This became known as the problem of time because it suggested that time and change simply do not exist.

"That's a theorem," says Linde. Yet "we are talking, and you are recording." The universe clearly does change—and yet the mathematics behind the equation seems to hold. So what's going on? Is our conversation now just an illusion?

DeWitt's interpretation provided a sort of way out. He noted that physicists never really look at the universe from the outside. Instead, they ask why we, situated within the universe, see something behaving one way or another. The presence of an observer is key. If you divide the universe into two, with one part containing an observer while the other portion is being observed, then the equations show that one side will start to evolve relative to the other. The clock will start to tick.

In the 1980s, Don Page, now at the University of Alberta, in Edmonton, Canada, and William Wootters, at Williams College, in Williamstown, Massachusetts, formalized the idea that time emerges under observation.[3] They imagined chopping the universe into two entangled pieces. (Entanglement, as we saw in Chapter 2, is a quantum link that permanently connects two quantum objects, so that changes made to one instantly affect its partner.) The key point was that to someone trapped in either of the two parts of the universe, time would appear to run within that section. But the evolution of both parts would counterbalance, so to a superobserver sitting outside of spacetime, the overall universe would be timeless.

In 2013, this esoteric idea got an experimental boost when Marco Genovese, a quantum physicist at the Istituto Nazionale di Ricerca Metrologica, in Turin, Italy, and colleagues created a model universe using just two particles, or photons, of laser light.[4] (Quite a sparse universe, but one that could be used to test the principle.) The team turned the photons into tiny clocks by making them rotate at a constant rate, ticking like little second hands. Each photon was also polarized, which means that it vibrated along its length in addition to rotating. Genovese's team then entangled the polarizations of the two photons, so if one was measured as moving up and down, its partner would instantaneously move side to side.

Each photon thus took on the role of one of the two entangled halves of the universe. By entangling themselves with one photon, the team tested the notion that within one "universe" you could see time pass. Physicists entangle themselves with systems simply by measuring the state of the system. So in this case, that involved measuring how far around the photon had rotated, and then measuring how far its partner had rotated relative to the first. Since rotation marked the passage of time, this served to show that time could be seen to evolve in one part of the "universe" if you observed it relative to another part.

To test whether a superobserver outside the universe, someone with a god's-eye view, would see time pass, the team repeated the experiment. But this time they were not allowed to become entangled with either photon, which meant that they could not measure how far either of them had rotated individually. Instead, as external observers, they restricted themselves to measuring only the joint state of the pair of photons—their joint polarizations. This allowed them only to confirm that the photons were constantly locked in an embrace, polarized in equal and opposite ways, which never changed. That is, from the outside, time did not pass for the universe as a whole.

"So as long as you do not have an observer, the arrow of time doesn't exist, and the paradox doesn't exist," says Linde. "As soon as you have an observer, the universe becomes alive." The arrow of time shoots forward and the cosmos evolves. "There is no absolute reality without mentioning what an observer would see."

If there were no internal observers, I ask, no conscious minds, no humans—or animals or dinosaurs—to observe what happens within the universe, would the universe change at all? Could any inanimate recording device "watch" the universe in the stead of a conscious being? Linde, from whom speech usually tumbles like a waterfall, slows and chooses his next words carefully. DeWitt, he says, argued that an inanimate recording device would suffice. "This was the only place where I think he was not brave enough," says Linde. For the recording device to change the universe, he argues, a consciousness would have to be observing the recorder. It would require a conscious mind to interpret the recordings and both register and understand that change had occurred to set the cosmic clock ticking.

So, I ask, what happened in the early universe before there was life in it? It sounds as though, in this formulation, the passage of time necessitates the presence of some meta-consciousness, observing everything. Could that meta-consciousness literally be God—and would that mean that God must be a part of the physical universe, an *internal* observer, so that He could see change elsewhere in the universe? I also think back to Antoine Suarez's claims (see Chapter 2) that quantum experiments prove that the story of the universe is being written from outside of space and time, a finding he felt complemented his belief in a God that exists in an immaterial realm. Following Suarez's line of thought offers an alternative identity for a deity: might God be an *external* superobserver, watching the universe from without and seeing no cosmic change, just a timeless portrait of the entire history of the cosmos laid out before Him?

Linde is not quite ready to address such questions yet. "People at this stage will say, 'And now I will take the Fifth,'" says Linde. He does add enigmatically that the best you can say is that the observations we make today are "consistent" with the idea that the universe existed before there was conscious life to experience it. I make a mental note to return to this later in our conversation.

We turn back to his student days, leading up to the development of the theory of inflation. I assume that this fascination with the underlying state of reality led him directly to study quantum cosmology—which in turn directly led him down the road to inflation. I was wrong. Linde tells me that his first attempt at coming up with his own theory involved developing a model to describe the "telepathic connection" between mother and child.

I blink, not sure that I heard that correctly. I'm wary of asking if I misheard him, in case he scoffs at me for even entertaining the thought that he was looking into something so, well, wacky. So I hedge, asking him to tell me more, hoping that he will spell out what he really meant so that I won't have to ask outright for any embarrassing clarifications.

"You didn't expect me to say this, did you?" he says, and laughs. His interest in telepathy stems, he tells me, from his early realization that the key to understanding reality is to understand consciousness. "An apple is something which may or may not exist, but my feeling of it is real. When I *feel* pain, it is real," Linde expounds. "The object that induces pain? I don't know if that is real or not."

Following this argument, he invented a theory to explain how a mother can simultaneously, almost telepathically, feel the pain that her child suffers, even though she has not been physically struck. For six months, he recalls, he was proud of his model, until he remembered reading that book on Einstein's special theory of relativity, which, as outlined back in Chapter 1, tells us that there is no such thing as objective simultaneity. Depending on how two observers are moving relative to each other through space, they can legitimately disagree on whether two events happened at the same time or whether one occurred before or after the other. Linde recalls being astonished by his premature celebrations. "If I did not know about the special theory of relativity, I would for the rest of my life be very proud of my theory. I would not be aware that it contradicts the rules of the game of physics," says Linde. "So then I decided that if I want to be a good philosopher or whatever, then I must learn the rules of the game again. So I became a physicist again."

Back at Moscow State University, Linde—like Guth—first turned his attention to the study of the building blocks of the universe, elementary particle physics. He asked his professors for advice on whether he should specialize in experimental physics, taking him to a particle accelerator, or in the theoretical aspects, coming up with calculations. The experimentalists told him to focus on experiments, since a bad theorist can lose touch with reality entirely, whereas a bad experimentalist will always at least be guided by nature when looking at the results of those experiments. The theorists told him, however, that a bad experimentalist is just one of hundreds of other bad experimentalists working on the same problems without understanding what they are doing, whereas a bad theorist working alone will at least come to grips with the laws of nature. "Somehow none of them considered the possibility that I may be a *good* physicist," Linde adds with a shrug.

Linde decided of his own accord that he had the potential to be a good theoretical physicist and went on to study cosmological phase transitions—the changes of the Higgs fields in the early universe that caused an initially unified force to split into the fundamental forces we know today, as we saw in Chapter 3—with his graduate advisor, David Kirzhnits, at the Lebedev Physical Institute. In 1972 they came up with a theory that was similar to, but less detailed than, the one that won Steven Weinberg and others the Nobel Prize a few years later.

It was a meeting at Lebedev that led to what Linde has referred to as the single most important thing to happen to him in life—his marriage to Renata Kallosh, a fellow physicist, who today collaborates with him on studying the implications of inflation. He recalls being struck by Kallosh, a "young, beautiful bright star of theoretical physics." Hampered by his own feelings of "incompetence"—an inferiority complex he says remains with him today—he judged her to be out of his league.

One day Linde found himself seated behind her at one of Kirzhnits's talks, and his advisor mentioned Linde's work.

"And who is that Linde?" he heard Kallosh ask.

"Ah, you ask a question," Linde recalls thinking at the time. "Your children will be [called] Linde!" He laughs sheepishly and adds that this is most likely his romanticized memory of the incident, and that, knowing himself back then, he probably would not even have dared to think like that.

Linde and Kallosh met again two years later at Lake Ladoga, where he sang songs to her and recited the words of Russian poets who had perished under Stalin and whose work was now forbidden, so he had learned it by heart. They were married a few months later; their first son, Dimitri, was born in 1975, and Alex, their second, arrived in 1978.

By 1978, Linde and colleagues had also come to an understanding similar to Guth's, as outlined in Chapter 3, that it was possible to create conditions in the early universe in which a special kind of field that pervaded the universe caused the cosmos to inflate at an exponential rate. Like Guth, they also noticed that there was a fundamental problem because they could not find a way that inflation would ever end in the universe—so the model, as set out, could not explain the state of the cosmos we see today. Crucially, unlike Guth, they did not publish what they saw as a flawed idea.

I ask Linde if now, with hindsight, he is kicking himself at allowing himself to be scooped by Guth. "No," he replies, sounding sincere. "I probably would do the same thing now." The key difference, Linde says, is that Guth recognized the potential for inflation to solve the flatness, horizon, and monopole problems and felt it was simply too good to give up on. Linde frankly admits that he himself had been blind to inflation's benefits. "Guth deserves a lot of credit for this."

At the end of his paper presenting inflation to the physics community, Guth noted his concerns that, according to his model, inflation would never have ended, writing: "I am publishing this paper in the hope that it will highlight the existence of this problem and encourage others to find some way to avoid the undesirable features of the inflationary scenario."[5] It would be Linde who would provide the answer. But first news of Guth's paper had to arrive in Moscow, which was no easy feat, given that Linde lived on the other side of the Iron Curtain separating the communist East from the capitalist West.

The path that academic papers took to reach Linde was slow. All journals and preprints arriving at the institute from the West had to be pre-approved for distribution by the institute's secretary, who would vet them for dangerous content—at least in theory. In practice, Linde splutters, "he never read them!" Instead, the papers languished on a sofa in the secretary's office, accumulating in ever-growing piles, until on occasion he deigned to pass one on.

So Linde only heard of Guth's paper later from a colleague, Lev Okun, who asked if he had heard about how inflation could solve the flatness problem. Linde replied, "No, I don't know anything about the paper, but I can tell you what it is about, and I can tell you why it does not work." And for half an hour, without ever seeing Guth's paper, he outlined Guth's mechanism and its major flaw.

As we saw in Chapter 3, Guth had visualized a hilly energy land-scape that the universe (filled with a Higgs-like field) could roll about in. High-energy regions (that is, areas with false vacuum) were depicted by tall dimpled hills. If the universe got stuck in the dip at the top of one of these hills, it would inflate at faster than light speed—exactly as Guth needed. The valleys marked the lowest-energy regions (the true vacuum), where the universe ultimately needed to end up. Once the universe was in the true-vacuum state, inflation would end, and the universe could get on with the job of evolving galactic structure and life.

So all Guth needed to do was carve out a route that the universe would likely spontaneously take from the top of a hill down to the bottom. He could even play master of the universe and shape the landscape's features in an effort to manipulate the universe into taking the perfect path. Being free to design a landscape with just the right topography to steer the universe might seem like a cheat—and in a way it is. There was, and is, no concrete evidence that such a special shaped field, with the right energy highs and lows, exists. But there is good reason to think that it might. The Higgs field was the first (and so far only) similar field to be experimentally confirmed to exist when in 2012 physicists at the Large Hadron Collider found the Higgs particle it produces. So it's not wildly unlikely that a similar cousin field, now somewhat unimaginatively dubbed the inflaton field, might exist. Unfortunately, its name remains the most definite thing we know about it; the inflaton field's exact shape, as Guth and Linde would realize, is up for grabs, and can be shifted in different versions of the theory depending on what you require the field to do.[6]

Guth had molded an inflaton field with just the right peaks and valleys for the universe to wander through, like an egg rolling down a hill at Easter. His trick was to construct an assault course that would trap the cosmic egg in a region of false vacuum long enough for the right amount of infla-tion to take place—enough to solve the flatness, horizon, and monopole problems. He did this by nestling the universe in the dimple at the top of a hill, where it would inflate. When that was done, the universe needed free passage home to the true vacuum at the bottom of the hill where inflation would end. If classical physics was the only thing at play, the cosmic egg would remain in the dimple at the top of the hill forever because it simply did not have enough energy to hurdle over the ridge. But quantum mechan-ics also raged in the early universe, allowing for some creative accounting in the egg's energy budget. As described in Chapter 2, quantum objects can tunnel through a wall that they cannot climb over. Guth was counting on the fact that, one way or the other, such quantum jiggery-pokery would get the cosmic egg to the other side of the ridge, where it could then roll down to the true vacuum.

But here was the problem that met Guth, and indeed Linde and his colleagues when they investigated a similar setup. The chances of tunneling happening decreases for bigger objects. That's why, even though it is a common enough process in the microscopic world, we humans never find ourselves suddenly quantum-tunneling through a brick wall and ending up, slightly puzzled, on the other side. When Guth did the math, he calculated that the chances of the entire universe tunneling in this way were slight— even with quantum mechanics waving its wand. Thinking of the universe as one solid, indivisible egg just made things too hard. Guth could not get Humpty Dumpty over the wall, so the whole theory went splat.

But there was a glimmer of hope. The odds of tunneling were much better if Guth broke the universe up into small patches. Then there was a reasonable chance that the patches could haphazardly tunnel through the wall one at a time. In this picture, bubbles of true vacuum popped out one by one in the false-vacuum sea that spanned the universe, like bubbles of gas in a pan of boiling water. What that meant for the actual cosmos was that inflation would suddenly end in those small patches where bubbles of space suddenly transitioned from false to true vacuum. Those cosmic patches would still continue to expand, but far more slowly than the rest of the still-inflating universe, which remained filled with false vacuum. All Guth had to do was to find a way to get these true-vacuum bubbles to merge again to rebuild one large universe.

Guth had imagined that this would be easy enough to engineer. He figured that soon after forming, the newly born bubbles of true vacuum would smash into each other, bursting their walls. Once all the bubbles had popped, the whole universe would be in the true-vacuum state and inflation would be done. This picture had another big selling point: each bubble collision would generate heat—providing enough energy for the universe to spontaneously create matter, and eventually atoms and molecules.

But things were not that simple. A closer look revealed that this would not work because each new true-vacuum bubble was embedded in the larger (false-vacuum) universe. Since that larger background universe would still be inflating, it would fling the true-vacuum bubble away rapidly, long before the next true-vacuum bubble could appear to collide with it. In other words, most of the universe would continue expanding at an alarm-ing rate, while inflation would have ended only in small, vastly separated areas of true vacuum that were scattered around and constantly being pulled farther apart from each other. Guth was left with an ever-inflating universe, forever blowing tiny bubbles.

Inflation, then, was too much of a good thing. It would not end every-where within the observable universe. If Guth's model was correct, it would mean that our universe today should still be inflating at breakneck speed. Matter would never have formed, let alone have clumped together to make stars and galaxies. It looked like inflation's metaphorical bubble had burst—ironically because its actual bubbles would not.

Guth's tribulations left Linde downcast. When he had pondered infla-tion with his colleagues, he had not realized its potential, so it was relatively easy to dismiss it as a failed theory. But Guth's paper shifted his perspective because it highlighted the many cosmological fruits that inflation could reap, if only it could be made to work. Inflation had come so close to solv-ing not just one but three cosmic conundrums, and with the demise of the theory Linde felt its loss on a somatic level, developing a duodenal ulcer, which today he puts down to his stress over the theory. "It was really emo-tionally disturbing that we have such a beautiful possibility to solve the problems of cosmology and it just plain doesn't work," he says.

Linde spent the next few months going in and out of the hospital with his ulcer, and it was while he was recovering in the summer of 1981 that inspiration struck. Guth had assumed that the false-vacuum peak had to be high and steep and that a patch of the universe would have to be stuck at the top long enough for inflation to happen, because as soon as it tunneled down to the true vacuum, inflation would end. But Linde had been running computer simulations of various tunneling scenarios and realized that the peak could be much flatter.

Linde's escape strategy from inflation's tyranny was to set up a low hill with a long, shallow slope. Then he returned to Guth's original plan of having the whole universe begin and end inflation at the same time, rather than being broken into patches, with inflation ending at different times in different patches. Linde also flattened the dimple at the top of Guth's high hill, which Guth had used as a place for the cosmic egg to temporarily nes-tle. That dip was not needed anymore because Linde's slope was so shal-low that the egg simply rolled along the top slowly, inflating all the while, before eventually plunging gently into the valley, where inflation ended.

For our universe, that meant that the whole cosmos went through inflation at the same time in its early history, and then the faster-than-light expansion halted everywhere across the cosmos simultaneously. Post-inflation, our entire universe would have continued to expand, but at a much more relaxed speed.

Linde's model could also explain how the universe heated up when it reached the true vacuum, so that it could create matter. His egg didn't stop dead when it reached the bottom of the hill. Instead, it rolled back and forth in the valley for a short while, and as it did so, the energy generated by these oscillations filled the universe with matter.

For a while Linde did not believe his own model because it seemed too easy. "I thought, 'My God, this is so simple, this is just trivial,'" says Linde. His first thought was that others must have found this solution already—and found an obvious reason why it must be wrong. At home late at night, afraid of waking his wife, Linde carried the phone into the bathroom and closed the door. Seated on the floor by the bathtub, he called one of his colleagues, Valery Rubakov, who had worked on the earlier, Russian version of inflation. Rubakov confirmed that, as far as he was aware, nobody else had made this leap. When the import of the discovery began to sink in, Linde woke up Kallosh—a move he describes as "almost a crime" given how tired she was from having cared for the two children all day. "Renata," he whispered, "it seems I know how the universe was born."

Now that he had the idea, Linde needed to disseminate it. But sending information out of Russia was almost as hard as breaking through one of Guth's bubble walls. All academic papers had to be accepted first in Russian journals; then authors had to request permission for the papers to be translated into English and submitted to a foreign journal. Approval was granted only if the material was deemed suitable for overseas ears. The process could take many months or even years.

But a few months later, while waiting for publication, Linde had a lucky break. There was a huge international physics conference being held in Moscow in October 1981, attended by Stephen Hawking, who had recently been appointed Lucasian Chair of Mathematics at the University of Cambridge—a position once held by Isaac Newton. At the conference Linde announced his "new inflation" model. It was received with excitement, with many participants offering to smuggle his paper out of the country—although by then he had received the required permission to publish abroad, so it was not necessary.

Hawking, however, was less impressed. The next day he gave his own talk. Since amyotrophic lateral sclerosis made speech difficult for Hawking, he would often ask his students to give his talk, interrupting to make clarifications as needed—and this time Linde was asked to translate into Russian. The talk was about Guth's inflation and its problems, and was progressing slowly, Linde recalls with a laugh: "Steve would say one word, the student would say one word, and then I would say one word." Eventually, because Linde was so familiar with the subject, he started to elaborate, adding his own thoughts: "So Steve says one word, the student says one word, and I talked for five minutes!" Finally Hawking mentioned that he had heard Linde's "interesting suggestion" for a solution a day earlier. Linde happily translated, filled with pride. Then Hawking added, "But this is wrong!"

Linde regards the half-hour that followed as the "most stupid" and "most embarrassing" situation of his life, as Hawking proceeded to tear his model to shreds in front of Linde's superiors while Linde was forced to stand there and translate. After Hawking had finished, Linde turned to the audience and said in Russian, "But I disagree!" and gave his rebuttal.

As they left the auditorium, Linde offered to explain his thoughts to Hawking, and the pair found an empty room, where Linde answered Hawking's criticisms for the next hour. But the pair had forgotten to alert anyone to what they were doing, leading the conference hosts to raise the alarm and set up a search for Hawking. "The whole institute was in a panic because suddenly this famous British scientist had disappeared and the next day there would be headlines about Hawking's kidnapping in Moscow," says Linde.

Eventually Hawking was found alive and well with Linde, and news of Linde's model trickled out to the West. Word spread to Guth, whose first thought—as he told me when we met at MIT—was that he had already tried to treat the universe as a single bubble and failed. "My first reaction, after reading the first sentence of the abstract, was that this must be nonsense," Guth recalled. So he approached the pretender's paper with skepticism, scrutinizing it for a mistake. But he couldn't find one. "Then, after I read the paper, I was convinced that it was a great idea, a way to solve the problem, and I was ecstatic, really," Guth says. "It was the first genuinely workable model of inflation."

That workability was about to be tested, however, as aficionados of inflation came together in Cambridge, England, at Hawking's invitation. Linde and Hawking's impromptu turn as fugitives had set up a deep friendship between the pair and renewed Hawking's interest in the theory, so in the summer of 1982 Hawking invited Guth, Linde, and others to a three-week workshop to flesh it out and to come up with some observational predictions. Guth felt so honored that he still keeps a copy of the invitation letter in his office.

On the surface, things seemed to be booming for inflation. It had grown in stature as two stateside physicists, Paul Steinhardt and Andreas Albrecht, both then at the University of Pennsylvania, also published a version of the new inflation model, in 1982, and Steinhardt was also at the Cambridge meeting.[7]

Hawking had set the goal of working out what kind of pattern inflation would imprint on the cosmic microwave background. As we saw in Chapter 1, this radiation, left over after the big bang, provides a snapshot of the temperature of light in the universe just about 380,000 years after inflation is thought to have taken place. As the universe dropped down to the true vacuum, the inflaton field would oscillate back and forth, generating quantum fluctuations. These, in turn, would leave gentle ripples on that temperature map, so there should be slight variations that astronomers could pick out. If the group could work out what form these bumps in the otherwise smooth background radiation should take, and if future measurements of these temperature patterns should indeed find those wiggles, it would be a huge boost for the theory.

These wrinkles in the temperature of light, caused by quantum fluctuations, would in turn reveal a corresponding underlying pattern in the density of matter in the universe. These uneven bumps were crucial because in regions where matter was slightly denser, there would be a slight boost in gravity, which would draw in even more matter. This clumping would thus be amplified, eventually leading to the massive conglomerations of stuff that would go on to constitute galaxies. The galaxies would be separated by vast tracts of nothingness that had started out in the infant cosmos as areas with an ever so slightly lower density of matter. In Linde's words: "We are children of these quantum fluctuations."

But before the workshop even began, Linde had his doubts. One of his colleagues who had worked deeply on the Russian version of inflation, Alexei Starobinsky, was now looking in detail at Linde's new inflation. The size of the density wobbles predicted by new inflation was inversely proportional to the speed at which the inflaton field, the "cosmic egg" in our analogy, rolled down the hill—the slower it rolled, the bigger these

bumps in the cosmic microwave background radiation. When Starobinsky crunched the numbers, the size of the density wobbles he found was
"waaaaaaaay too large," says Linde. The density wobbles would have
pulled matter together so tightly that essentially all the matter in the universe would have collapsed into black holes, with no chance for planets or
people to form.

Hawking, meanwhile, had performed a calculation similar to Starobinsky's, but he had somehow managed to get an entirely different answer that
perfectly matched the level of quantum fluctuations needed to give rise to
structure in our universe. Who was right? With the help of colleagues, Guth
tried to reproduce Hawking's calculations before coming to the meeting,
finishing them up in Cambridge in his hotel the night before the meeting
started. His answers matched Starobinsky's pessimistic version, however.
Pretty much all of the other participants agreed, having checked the results
using different mathematical methods. They all concurred: new inflation
could not explain our universe. All, that was, except Hawking, who dug in
and defended his calculation.

The younger men, intimidated by the stature of their host, argued that
evening over who should step up and declare that Hawking was wrong.
The next morning, as Hawking approached the end of his talk, Guth says
that the others were primed to jump up and object. But then Hawking simply announced an answer in agreement with theirs—never mentioning that
he'd ever thought differently, or that he had changed his mind.

Despite the downturn in mood, Linde regards the meeting as one of the
most productive of his life, with minds coming together to think through
the moment (he claps his hands for emphasis) around inflation's incep-
tion. Nonetheless, he left knowing that new inflation, as the model was
set out, would not work either. It was possible to contrive some very oddly
shaped energy landscapes for the inflaton field that might give the right
level of density perturbations—but that seemed arbitrary and, frankly, a
bit desperate.

As Linde worked on a remedy, he sank into a period of deep reflection
that caused him to review quantum cosmology—in particular, the ideas
about the wavefunction of the universe and the problem of time set out by
the Wheeler-DeWitt equation—in a more contemplative light. Though he
had not been raised in any religious background, his physics had begun to
shape his philosophy. It was no coincidence that the year he invented new
inflation, allowing him to read the blueprints of the universe's creation, he

also began studying ancient Indian philosophy in depth, particularly the Advaita Vedanta. He stresses that he is no expert, but his readings made him think about the different ways that God could be conceived: at one extreme "as a man with a beard" or as a "powerful force of nature," but at the other, more abstract end, outlined in these Indian teachings, as "perfection" and as "absolute."

God, as absolute perfection, the ancient Indian scholars proclaimed, is unchanging. After all, if perfection evolved into something else, then either the initial state or the final state would have to be less than perfect. This brought Linde back to thinking about the Wheeler-DeWitt equation, which showed that the universe, when viewed from outside, cannot evolve. "You think about the wavefunction of the universe, which is absolute perfection, which does not depend on time, which embeds everything at all—everything including us observers," says Linde. The philosophical literature he was reading from many thousands of years back seemed to resonate with the physics of quantum cosmology today. They spoke of human consciousness "cutting itself out of God," says Linde. It reminds me of my discussion with Abhay Ashtekar, recounted in Chapter 2, and his notion that individual human consciousnesses are excitations out of a larger communal field of consciousness, but in the texts that Linde refers to, this wider consciousness is explicitly identified with God. Furthermore, Linde recalls, his readings in ancient philosophy described that separation from God as enabling humans to experience the passing of time.

Linde quietly, and nervously, describes his own personal philosophy. "This duality between you and the universe is like a part of the whole package," he says. "So maybe when you die, you become part of the whole, and then maybe you get separated again, in a different way." He grins, seeming a bit abashed. "All this sounds very, very religious. All very, very superscientific," he chuckles. "If you want to have my reputation just destroyed, I guess you have enough material," he says, pointing at my notes.

I'm slightly surprised by his openness here. In fact, a few days earlier I had spoken to someone who described Linde as one of the friendliest and most open-minded atheists he had met. While Linde does not identify with any organized religion, the words I'm hearing today are not those commonly heard from strident nonbelievers. I double-check whether what he's telling me is something that he holds dear on a personal level or just a flight of whimsy. "No, I don't think it is just intellectual musing—I think that this is an integral part of what I am," Linde says deliberately.

Linde explains that he will not readily use the word "God" when describing his beliefs, because the word has been misused and abused and can be wielded as a weapon to inspire people to carry out horrible acts. "When you try to use this word, you are kind of shaking," says Linde. His speech has now slowed, with almost a second's pause between each carefully chosen word. By way of analogy, he talks of those who studied radioactivity when it was first discovered, handling hazardous material while unaware of its dangers. "When people are thinking about something which is potentially powerful and important but not yet understood, they were sitting in close proximity of decaying materials, without understanding that this is what they do not see, that it can actually kill them," says Linde. "So these things are better approached with utmost care and with worry that whatever you say can be easily misinterpreted."

His words describe the care he takes when talking about God, religion, and his understandings of the parallels between quantum cosmology and Indian philosophy. But his colleagues, and soon he too, would go several steps further. Linde's attempts to fix inflation again would become entangled with the research of another Soviet, Alex Vilenkin, who was about to devise a way that our universe could have been created from nothing—and to stumble on the possibility that our universe is only one of many parallel cosmoses in an infinite multiverse. Both ideas would work their way into the blueprints for making a universe in the lab.

5

Universe to Multiverse

Alexander Vilenkin is a man who has lived multiple lives: as a soldier, as a zookeeper, and as a physicist. His story could probably fill a number of biographies. So perhaps it's not surprising that this shy Ukrainian should come up with a theory of a multiverse of parallel universes, containing an infinite number of clones of each of us.

En route to this discovery, Vilenkin also studied the creation of our universe from "nothing"—a model that some have argued stands as proof that there's no need for a God (or even a physicist hacker) to have made our cosmos. Vilenkin did not set out to perform research at the boundary between science and religion, but he has somehow become embroiled in the debate between the two. "I don't enjoy being the center of attention, and I certainly don't look for a fight," says Vilenkin quietly. "But I did get into some controversial fields." Vilenkin credits his discoveries to his "obsession" with the unobservable. It's this trait that led him to come up with a series of ideas that will go into the universe builder's user guide.

I've known Vilenkin, whom I've come to visit at his office at Tufts University, in Medford, Massachusetts, for some years. You can pick him out at meetings because he sits in lectures wearing shades. He has a delicate dancer's frame and a soft voice. He speaks slowly and carefully and with a natural humor that's delivered deadpan, so for a moment you cannot be sure he is joking, till he giggles.

Vilenkin's first brush with physics was in elementary school, in the town of Kharkiv. He was inspired to study science by a charismatic math teacher, who fascinated students both because they knew he'd had an "interesting" past during the Russian Revolution and because of his teaching. People were not encouraged to speak about their pre-revolution lives in the Soviet Union—and certainly not to schoolkids—so while the children did not know the details, they were enthralled by the romance.

Vilenkin was encouraged to apply to a specialized mathematics high school, where he excelled in physics. Having read popular accounts of how Einstein predicted that time could stretch and length could contract, depending on how you were moving, he and a friend decided to learn the subject from Arthur Eddington's book *The Mathematical Theory of Relativity*. The book ended with a discussion of cosmology that captured Vilenkin's imagination. "I was amazed that you can say something about the beginning of the universe," he recalls. "I was fascinated with the beginning. Something very mysterious obviously happened then. And even though we have the big bang cosmology, which brings us pretty close to the beginning, it doesn't tell us what actually happened." Once Vilenkin realized there was a field of study dedicated to finding out the answer, he knew that would be his profession.

These were unfashionable questions to be thinking about, however. The Soviet Union was a "militantly secular" country, and questions about the beginning took you to a place that went beyond science, at a time when cosmology had barely established itself as a science rather than as a branch of metaphysics. "In some sense, physicists were reluctant to talk about these things. They were kind of taboo subjects—regarded as a totally pointless thing that people do when they have no better things to do," Vilenkin recalls. "Philosophy is a bad word in the physics world, so if someone tells you that you work at philosophy, it is not a compliment."

At Kharkiv University, Vilenkin appeared to be on track for a stellar career in physics, though his interest in esoteric ideas about the origins of the universe troubled some of his professors, who tried to encourage him to take up a more practical branch of physics. "All this relativity is just play," Vilenkin remembers being told. "You have to do some real physics."

Heeding that advice, Vilenkin officially specialized in condensed-matter physics—the study of the interactions within materials, which forms the basis of superconductivity and has many applications for engineering. In his spare time he continued to study relativity with a cosmologist at the university. But when the time came for him to graduate, things started to fall apart.

In part this was because Vilenkin was ethnically Jewish, though he had little experience of Judaism as a religion. Neither his parents nor his grandparents observed the faith. "I had no idea about religion," says Vilenkin. What had once been the only synagogue in the city had long since been turned into a sports center, where Vilenkin fenced. Today, he says, if he had to classify his beliefs, he would call himself agnostic. But at the time his Jewish ethnicity became a hurdle to his career. There was no unemployment in the Soviet Union, and fresh graduates on an academic track were usually assigned a place to continue their education. Vilenkin was officially directed to Kiev University to work with an expert in general relativity. But the professor to whom he was assigned, a man named Petrov, was a notorious anti-Semite. Vilenkin went to meet with Petrov, and although the professor would not say explicitly that he did not accept Jewish students, "his secretary told me, 'Don't expect an invitation from us—it's not going to happen.'"

Though Vilenkin did not have any religious beliefs, he did subscribe to communist ideals in his youth, voraciously reading Marx and Lenin and attempting to make sense of their writings, "until the embarrassingly late age of about nineteen or twenty," he says. (He notes with a smile that this brief flirtation with communism got him into trouble when he later applied for citizenship in the United States and had to admit that he had believed in communism until the age of twenty. His wife, who had not held such beliefs, got her citizenship immediately, but Vilenkin's was delayed for six years—this period evidently being deemed enough to balance out his youthful transgression.) But Vilenkin became increasingly disillusioned with the Soviet system at university, where another obstacle was placed in his path.

Despite his having been rejected by Kiev University, there were plenty of other professors in Kharkiv who knew and liked Vilenkin and helped him to secure a number of good offers. Yet each time he attempted to take one up, at the last moment the offer would be retracted. Vilenkin realized he had been flagged by the Soviet secret service, the KGB, as an unsavory character. "They don't tell you that you are blacklisted, they don't send you a letter, and you never know, but I have some guesses why," Vilenkin says. "I wasn't any freedom fighter, or anything like that, but you could do reasonable ordinary things—make a joke in the wrong company—and that's enough to get you in trouble."

It was then his life took a somewhat unconventional detour, at least for a physicist. After a year of trying and failing to get into grad school, it was clear to Vilenkin that academia in the Soviet Union was not an option. He was drafted into the army and spent a year as a soldier before finding work teaching mature learners at an evening school. Vilenkin relished teaching. Though the students did not want to learn rigorous physics, they were enthusiastic and fascinated by how rockets work and how you might travel to other planets. The trouble was, Vilenkin's duties extended beyond the classroom. If students did not show up for class, it was his job to chase them down, which usually entailed knocking on the door of an angry, drunk truant and dragging that person bodily to the lesson. So when Vilenkin heard of a job opening at Kharkiv State Zoo, the prospect of handling animals rather than people seemed eminently more appealing.

Vilenkin's first duty was to guard the zoo's kiosk, which sold alcohol. He managed to get the job because he could provide letters of recommendation vouching that he was not a drunkard and thus would be unlikely to deplete the booty he was meant to be protecting. Soon he made friends with the night watchmen guarding the animals, and got a joint position working with them. By now, Vilenkin was married with a small daughter, so making a double salary was a huge boost.

It was a surreal time for Vilenkin. His physics friends would visit him at night at the zoo, which he patrolled. "There was a rifle, which I didn't actually know how to shoot," says Vilenkin. "I just hoped the other guard did." The zoo was fairly easy to break into—you simply had to clamber over a fence—and a swan and an eagle did go missing, "but fortunately not on my shift," adds Vilenkin. This was good because Vilenkin's attention was not wholly focused on the animals. With his physics guests at the zoo, he chatted about cosmology and his other projects, and read physics papers. One of his publications, in which he combined quantum effects with general relativity to make predictions for how a particle's energy may change, lists his academic affiliation as "Kharkiv State Zoo." (None of the animals were acknowledged in the paper as sources of inspiration, however.)

Looking back, he acknowledges that the experience has supplied him with a more colorful set of anecdotes than many of his physics colleagues can produce. But back then, it was not so amusing. "When you don't really see how this is going to lead to anything, at the time it wasn't easy," says Vilenkin. "The important thing is I tried to keep in shape in terms of my physics."

There was a way out of the country for Jews, though, if they could furnish an invitation letter from a family member in Israel. Vilenkin had no such family member, but a system was in place in which Israeli citizens would pretend to be distant relatives to try to help people out of the Soviet Union. It took a year for Vilenkin and his young family to get permission to leave. Within days of Vilenkin's family making the application to the Soviet authorities, his wife's parents lost their jobs, and he believes it was because they gave their daughter their blessing to emigrate. Vilenkin's own father was pushed into early retirement from a university position soon after.

Upon leaving Russia, in 1976, Vilenkin spent a few months in Italy, where he applied for a teaching assistantship that was being advertised at the State University of New York at Buffalo, bringing him to the United States. The department there did not do much in the way of cosmology, so Vilenkin threw himself into a PhD on the physics of polymers—graduating in just a year—before moving into a postdoc in Cleveland, where he studied thermoelectricity, working out how electric voltages can be generated from changes in temperature.

Vilenkin was happy to be out of the Soviet Union; he had assumed that, as a new immigrant, he would be washing dishes, but instead he had a good job. "Every time I woke up, I couldn't believe my luck at being here," he says. This dampened any frustration he felt at not being able to sink his teeth into cosmology immediately. It also, inadvertently, may have been an advantage. He was working on cosmology in isolation as a hobby and so had relatively little notion of what topics were most popular—and, in particular, which questions he should avoid asking.

In 1978, Vilenkin came to Tufts, where he has been ever since. It was only a short while later that Alan Guth developed inflation (see Chapter 3). Vilenkin heard him speak at Harvard in 1980, while Guth was touring to promote the theory (and find a job), and he was immediately taken with the fact that inflation could solve so many puzzles about the universe that many had assumed could never be explained. "It was amazing because he addressed some questions that were thought to be 'beyond science,'" says Vilenkin. "I started thinking about it and actually working on it pretty much right away after I heard the talk."

The particular claim that caught Vilenkin's attention—which most other physicists had overlooked—was that inflation could conjure matter in the universe, seemingly from nothing. Since then Guth has put this somewhat cheekily, saying, "In the context of inflationary cosmology, it is fair to say that the universe is the ultimate free lunch."[1] But Vilenkin was not so easily convinced. "It's not quite a free lunch because you still need a little speck of space, or false vacuum, to inflate," he explains. You don't get something for absolutely nothing. "I was thinking, you know, that of course it's very little, but still, where did that come from?"

Vilenkin imagined what might happen to an infant universe containing matter and also vacuum energy. (Recall from Chapter 3 that even apparently empty space has some energy; this vacuum energy provides an outward push, while the gravity between matter in the universe will try to pull the cosmos back inward.) He figured that a tiny universe in which matter was densely packed, providing a huge inward pull, would probably recollapse before it had a chance to really grow. A bigger universe, on the other hand, in which matter was more diluted, would still have the outward thrust provided by its vacuum energy and so would be able to expand. There must also be a critical radius between these two extremes of collapse and inflation.[2]

The question was, if your universe starts out with a radius that's less than this critical value, was there any way it could be transformed into a universe with a radius larger than the critical value without actually having to go through the hard work of physically expanding? It's analogous to asking if a quantum particle could ever traverse a seemingly insurmountable energy barrier and find itself on the other side. The answer, for quantum objects, is yes—quantum tunneling, as we saw in Chapter 2, sometimes lets you cheat and hop over hurdles that you should not be able to overcome. And since Vilenkin was thinking about just a teeny speck of universe, he figured those quantum laws should still apply.[3]

So, just as you can use quantum equations to predict how likely a particle is to tunnel across a high energy barrier, Vilenkin calculated the probability that an initially tiny bubble of false vacuum would quantum-tunnel to a larger one and inflate into an adult universe before it had time to collapse. In practice, that would mean that the baby cosmos started with a small radius and grew gradually, but then all of a sudden—as it quantum-tunneled—it instantaneously transformed into a universe with a much bigger radius, without needing to pass through the intermediate sizes. He found that this probability was small but non-zero, so it could happen. But then, as he shrank the size of the initial bubble, "I noticed an interesting thing: the probability did not really vanish even when the size of the universe went to zero," says Vilenkin. "So when I basically remove this initial universe, I still have this probability of tunneling—it's not clear from what!"

It took some time to sink in, but, says Vilenkin, "in the end it kind of hit me that this universe was created out of nothing." I want to push him on this a little. His paper, published in 1982, was called "Creation of Universes from Nothing."[4] But there is a question of how to interpret what he found. It sounds odd to take away all the starting ingredients in your calculation—all space and time and matter—and yet still interpret your end result, the new universe, in a spacetime context. "It seems crazy. Did you think it was crazy?" I have to ask.

Vilenkin chuckles under his breath. "I said this in the paper in so many words," he admits. "Its interpretation, of course, was not clear at all." But for Vilenkin, at least, it served to put an end to the infinite regress of questions about what happened before inflation, or what happened before the big bang. "How can we end this? Here the answer appears to be, there was nothing," Vilenkin argues. "And I don't think there were any answers to that question before."

But what is "nothing"? There may be no pieces on the playing board, and in fact no board itself if you take away space and time, but for Vilenkin's idea to work, the rules of the game must still be in place. This situation of spacelessness and timelessness, for want of better terminology, must still obey quantum laws. As Guth asked in Chapter 3, where did those laws of physics come from?

"Well, 'nothing,' of course, I had in quotation marks, because you still need the laws of physics to govern this process," agrees Vilenkin. "But it was nothing in the sense that there is no macroscopic spacetime and no matter."

It makes my head hurt to think that there can be laws of physics that govern how reality should behave before there was ever a reality there to govern. But then, perhaps that in itself is some kind of definition of God. Vilenkin points out that while it's strange that these laws hold sway over nothing, it's no more baffling than holding sway over the world around us now. "The laws of physics have power when there is something, and it is not clear how they do that either," he remarks.

But the idea—indeed, the very fact that he dared think about the puzzle of how to create a universe from nothing—was controversial. Soon after publishing the paper, he gave a talk at Harvard, where fellow cosmologist Lawrence Krauss told him that he was "amazed" Vilenkin had been able to give such a talk "and survive." Krauss, it turned out, had had similar thoughts, but had held them back for fear of ridicule.

For Vilenkin, the key to bringing discussions of the beginning of the universe out of philosophy and into science is the fact that he uses mathematics to back up what might otherwise be speculation. It's led him to think about whether there may be a way to test this idea (in a slightly less ambitious way than making a universe in the lab). It seems that there could be consequences to his theory that, if he's correct, have already been observed.

Vilenkin's argument goes like this: A universe that spontaneously pops into being, having tunneled into existence from "nothing," would be born in a particular quantum state, with a certain wavefunction. Each possible wavefunction would create a different vacuum in the early universe, leading to different inflationary perturbations, which in turn trace out different patterns on the cosmic microwave background. Cosmologists have assumed that the initial vacuum was what's called a Bunch-Davies state (one of the most basic kinds of possible vacuums allowed in theory; the details of the vacuum are not too important here, though), and this appears to match observations. But they have no fundamental explanation for why the universe would be born that way; it's just a nice and simple vacuum to work with. Handily, though, it is also the vacuum predicted to appear if the universe is created by Vilenkin's method. This does not stand as a prediction of something new to search for in order to confirm the theory, as this vacuum state is already assumed to be correct. But it does support Vilenkin's idea.[5]

Having sorted out, at least to his satisfaction, the beginning of the universe, Vilenkin then turned to the question of what lies beyond our visible universe—the regions so far away that light has not yet had time to travel from there to Earth. "You can see that I have this attraction to unobservable things," Vilenkin says. In particular, he turned to Andrei Linde's new inflation. Recall from Chapters 3 and 4 that inflation's architects have proposed that the universe is filled with an inflaton field that can gain and lose energy, as though it was ascending and descending a hill. When the field is temporarily lodged in a high-energy false-vacuum state, the universe inflates, expanding faster than the speed of light. Inflation ends when the field drops into the low-energy true vacuum; at this point, the cosmos would still continue to expand, but at a much more leisurely pace. In Linde's model the false vacuum is fashioned as the top of a flattened hill, while the true vacuum lies in the valley below. Vilenkin wanted to investigate what would happen if different patches of the universe dropped into the true vacuum at slightly different times.

In Chapter 4, I introduced a somewhat flippant analogy comparing Guth's inflating cosmic egg to Humpty Dumpty. At the risk of stretching the joke too far, I will now call on the egg's equally stability-challenged chums Jack and Jill to stand in for different patches of the universe. Vilenkin set both Jack and Jill atop Linde's flattened false-vacuum hill and then watched them tumble one after the other down to the true vacuum. In the real non-nursery-rhyme universe, this corresponds to the rate of expansion dramatically slowing in some patches of the cosmos (when they reach the true vacuum and inflation ends), while the rest of the cosmos (still in the false-vacuum state) carries on rapidly inflating.

So, as before, as long as the patches of the universe they represented remained at the top of the hill, both Jack and Jill would inflate. When Jack and Jill fell down the hill and hit the valley, however, inflation in their patch would end. But what, Vilenkin wondered, would make Jack and Jill topple in the first place, assuming they had started off reasonably well balanced on the peak?

The answer was the spontaneous quantum fluctuations that are responsible for producing pairs of virtual particles in the vacuum, discussed in Chapter 2. Each time a pair blipped into momentary existence, it would give a slight nudge to anyone (or any patch of universe) standing on the hill. No single quantum shove would be enough to knock Jack, say, right off the hilltop. But he would gradually and randomly be bounced around the surface of the hilltop, sometimes veering toward the edge and sometimes being knocked back toward the center. In the meantime, between each push, Jack's patch of the universe would double in size, thanks to the fact that inflation was still operating on him. Only after a series of shoves brought him perilously close to the edge would he topple over, plummeting down into the true vacuum, ending inflation in his patch.

But what about Jill, Jack's neighboring patch of universe? Would she really come tumbling after? If Jill is originally close to Jack—that is, if two patches of space start out fairly close to each other at the onset of inflation—there's a good chance that, yes, she will fall alongside him and share his downfall. Within our universe, that means that these two patches will have followed the same inflationary paths and have a shared cosmic history. All of the universe that lies within our observable horizon today corresponds to patches that started out like that, in very close proximity.[6]

But what if Jill was farther away from Jack at the onset of inflation? In that case, the series of quantum nudges that tipped Jack over the edge might not be enough to unbalance Jill. So while inflation may stop in Jack's region of the universe, it may well continue in Jill's. Since the region of the universe represented by Jill is still inflating—growing in size at an exponential rate—the region once entirely described by Jill will need to be described by exponentially more people: Jane, Jessica, Jamilla, and so on. Some will follow Jack quickly down the hill, while others will take longer. The upshot of this is that, as Vilenkin realized, inflation will not end all over the universe at the same time. Each time one patch falls down (Jack, Jessica . . .), inflation in that pocket of the universe will end. Each will become a bubble of true vacuum in a still-inflating false-vacuum background.

This picture may look familiar; it was the one that Guth first tried to use to describe his inflationary model in the 1970s, described in Chapter 4. He had also divided his inflating universe into patches, and watched inflation end in each patch separately. His hope had been that these patches would reunite to form one true-vacuum-filled universe at the end of the process, but then he rejected the model precisely because he could not find a way to make all the pieces join up again to form our cosmos. Vilenkin's insight was that the pieces should not join up again. Instead, each patch of newly made true vacuum could be thought of as its own cosmos, floating in the still-inflating false-vacuum background. Just as Guth had seen, each true-vacuum piece will have been sent zooming away, faster than the speed of light, by the inflating background. Those pieces—Jack, Jill, Jane, and so on—would speed away from each other so quickly that they would lose any hope of ever being able to communicate with each other again.

In essence, each piece becomes its own bubble universe, divorced from its siblings by the larger inflating multiverse. Bubble universe creation continues forever, and inflation is eternal.

Vilenkin was not completely sure of the import of what he had found. "I was pretty fascinated, but when I tried to discuss it with some physicists, they were not particularly interested in such things," he says. One issue was that, by their very definition, these extra universes lie beyond our observable reach and so, seemingly, could never be detected. So for some, there seemed little value in musing about their existence. Disheartened, Vilenkin ended up sticking his ideas about eternal inflation onto the end of a paper about something else, without much of an extended discussion and little fanfare. He published his work in 1983, in *Physical Review D*, but it failed to grab much attention.[7]

Vilenkin also broached the topic with inflationary guru Guth, with whom he has become good friends. Vilenkin is now aware of Guth's habit of falling asleep during lecturers and physics discussions—but he didn't know that when he first visited Guth's office to outline his ideas, and so when he saw his idol snoozing, he took it as an indictment of his work. As a result, Vilenkin set aside the multiverse for the next decade.

Guth himself doesn't recall falling asleep in front of Vilenkin, though, he chuckles, he would probably only remember waking up. While he meant no disrespect to Vilenkin, "certainly the story of my falling asleep is at least morally valid," Guth admits, adding that it more or less summed up his attitude toward the notion of eternal inflation at that time. Guth accepted that inflation probably did go on forever, creating new bubble universes, just as Vilenkin was suggesting; he just did not see why anybody should care. At that point, it seemed that these endlessly inflating bubbles would have the exact same laws of physics as our own, and if that was the case, then why bother investigating? It was really just a slightly more exotic statement of the fact that the universe extended beyond our observable horizon.

It would be almost two more decades before advances in string theory made Guth pay more attention to the multiverse because that theory suggests that each universe could have very different properties, as we'll see in Chapter 7. But in the meantime, Linde was independently taking up the cause.

By 1983, Linde had lost faith in new inflation because calculations by himself and others had shown that it failed to produce the right level of quantum fluctuations to explain galactic structure. The problem had been highlighted the year before at a workshop held in Cambridge, as we saw in Chapter 4. But while others tried to tweak his theory to make it fit with observations, Linde was not optimistic. "There was a period of about a year where gradually it started sinking down that we really have a problem," Linde recalls. What inflation needed, he says, was a radical overhaul.

One of Linde's concerns was that from the outset inflation was kind of "tricky." Guth had come up with one energy landscape—a conveniently shaped hill with a dimple on top, defined by the inflaton field—that he thought should work, but it didn't. Guth had offered no fundamental explanation for why the energy landscape took that particular form, however.

After that, Linde himself had fashioned a shallow hill with a flat slope, but again, the same question could be asked: why did the energy landscape take this form and not some other that was less conducive to inflation? In the nursery rhyme, Jack and Jill climbed up the hill to fetch a pail of water, but with inflation, there was no clear reason why the universe ended up at the top of the hill—or, indeed, why the hill was there in the first place. In other words, there was no real explanation for why the inflaton field that pervaded the universe found itself in a high-energy false-vacuum state, enabling inflation to begin.

Linde's insight was that if he couldn't find a convincing way to make Jack and Jill go to the mountain, then he would just have to make the mountain come to Jack and Jill, spontaneously arising under their feet. His new model was altogether more anarchic, allowing the energy in different patches of the universe to suddenly shift dramatically and at random. Thus Linde turned his previously smooth, hilly energy landscape into a constantly undulating terrain, where peaks could turn into troughs and back again.

And so Linde's theory of chaotic inflation was born. Wait long enough and any small patch of the universe could, and most likely at some point would, suddenly find itself stranded in a false vacuum and start to inflate. The universe did not even have to find itself at the top of the hilltop; it only needed to be midway up a large enough hill such that it would take a long time to reach the bottom, giving enough time for sufficient inflation to happen on the way down.

"For me, this was the moment when I thought, 'Well, inflation's not a trick anymore,'" says Linde. "This was a signal that we were actually on the right track." Linde had developed the theory of chaotic inflation while in Russia, but the realization that it could provide an easy explanation for why the universe—or at least some portion of it—ended up in a false-vacuum state, triggering inflation, hit Linde as he was flying to the United States for the first time, to attend a conference on Shelter Island, just off the eastern tip of New York's Long Island. "How I was allowed to go there by Russian authorities, I have no idea," says Linde. It was a time of heightened tension between East and West, and many of Linde's colleagues had been refused permission to travel to the same meeting.

Linde himself had to take a circuitous route. There were no direct flights between Russia and the United States, so Linde had to stop in Canada. It was on the flight from Canada that he outlined the talk he was about to give, introducing chaotic inflation to the world. But just when the airplane reached the airspace over New York, it was forced to turn back because of storms. "I thought, 'It seems that God does not want people to know about chaotic inflation!'" Linde recalls. "I could have expected anything other than this mystical thing—the plane turning above New York and returning me back." God may well have heeded his lamentations, because after reaching Canada again, the plane switched course once more, back to New York.

At the conference, Linde reported his discovery. It ruffled feathers because it was so ideologically different from the style of the previous solution, which rested on precisely defined conditions in the early universe. "In this chaotic inflation story, you start with whatever garbage you have, and some parts of the universe, which are in a good position, they become huge," says Linde. "And those parts which cannot lead to inflation, they just don't expand, but who cares about them?"

Hawking, who had been persuaded to accept Linde's earlier theory of new inflation, understood and backed chaotic inflation immediately. With this seal of approval, Linde felt his job was now done. Under chaotic inflation, pretty much any underlying energy landscape could and would give rise to inflation—and some of those conditions should produce the correct level of quantum fluctuations to match the density perturbations needed to give rise to large-scale structure.

For the next two years, Linde comforted himself that everything had been sorted out. "Ideologically, everything was clear. The theory was primitively simple," he says. "I thought, 'It's time to stop and write a book.'" Linde did indeed sit down to document his theory in an academic book for physicists, a process that, he whispers to me, he hated: telling the same old story he had spoken about before, and using scissors and glue to literally cut and paste in corrections whenever he'd made a mistake.

Linde fell into something of a funk—which inadvertently would lead to his next breakthrough about the multiverse. It was the year of perestroika, and Soviet leader Mikhail Gorbachev removed the archaic system for publishing papers, to try to facilitate communication with the West. Unfortunately, he did not replace it with anything else, so once again Russians were left without an international outlet. Linde became increasingly despondent, unable to work, and feeling like a "cow wanting to be milked—it was painful."

He and Kallosh had gathered enough money to buy a car, so Linde, then in his forties, took driving lessons on Moscow's icy winter roads.

"This was an extremely humiliating experience because your instincts at that time, when you are not young, are all wrong, and the cars were primitive." Linde recalls his instructor, in frustration, using expressions that he is too polite to translate.

Linde fell into a state that he now recognizes as depression, and spent days in bed, feeling awful and too weak to walk or even stand up, though he was told by doctors that there was nothing physically wrong with him. Then he received a phone call from the secretary of the Lebedev Institute, where he worked, informing him that he had been picked to travel to Italy to give a series of popular lectures on astronomy to the public. Linde was reluctant to go. Aside from his health, academics were only allowed to leave the country at most once a year, and so he did not want to waste his trip on a popular science tour rather than on an academic conference. He informed the secretary that he was too sick to go, furnishing a certificate from the head of his group at the institute saying that he should not have to leave.

But getting out of the trip would not be so easy. The secretary called back saying that the first lecture was a month away, and so Linde should have recovered by then—unless of course, the secretary said pointedly, Linde was saying that he would never be well enough to go abroad again. "And then I understood this really smells dangerous," says Linde. If he was not careful, he would be barred from travel for life.

Linde forced himself out of bed and took a taxi to the hospital—not a minor expense—where he went through a week's worth of tests in a day, eventually procuring a certificate saying that he was in good physical health. The effort was draining, however. He spent the weekend back in bed, dreading what was to come. On Monday he made "another financial decision" to hire a taxi to get to the institute, where he took care of all the paperwork necessary to get permission to leave. "I signed everything in one day, which typically takes a month and a half, but I was really scared, so it was not a joking matter."

Linde handed the documentation to the secretary personally, who informed him that the Italian organizers of the tour wanted him to supply them with the text of his talk, which they would publish and distribute to all the citizens of Rome, by the next day. "Look, I was barely able to stand up," Linde says, shaking his head at the memory. It seemed an impudence, and yet Linde realized it also afforded him a unique opportunity to write an academic paper and have it sent out of the country immediately for distribution in the West, via diplomatic mail, without the usual checks and censorship. He was determined to make the most of it, having spent months in the doldrums, by thinking up something new. "What can I invent in half an hour, so I can print it tomorrow? What can I invent? It took hold of my head," Linde says. Half an hour later, he had converted his theory of chaotic inflation into eternal chaotic inflation.

Linde, like Vilenkin, had realized that different patches in the universe could start inflating at any time, and that inflation would always be happening somewhere. His shifting, chaotic energy landscape continually raised different patches of the universe at different times, depositing them on false-vacuum slopes and then leaving them to roll down to the true vacuum, inflating on the way. Inflation would be eternal. "When I found eternal inflation, it was like a moment of absolute clarity," Linde recalls. "This was one of the strongest emotional shocks of my life."

***

Both Linde and Vilenkin, then, had found that the birth of the universe no longer marked *the* beginning, but only *a* beginning. Our universe, in these versions of inflation theory, was no longer a unique phenomenon, but merely one bubble in a roiling froth, each undergoing its own cycle of birth, expansion, and possible collapse.

Humanity had been knocked off its throne at the center of the universe some 450 years earlier by the Copernican revolution, which stated that Earth revolves around the Sun, rather than vice versa. Now, it seemed, our universe may not even be unique. The idea has proved controversial— as we'll see in Chapter 7, when the multiverse goes head-to-head with God. For Vilenkin and Linde, though, the simple fact that there are other universes out there was no stranger than, in Linde's words, understanding that Earth is huge and that different countries exist around the globe.

But, fast-forwarding momentarily, there was one more revelation to come from Vilenkin, one that he describes as "disturbing" because it struck at the heart of his sense of meaning in life and personal identity. That was his realization that his work on the multiverse suggests that these parallel universes could give rise to an infinite number of copies of each of us.

The reasoning that led Vilenkin to posit the existence of these clones is often simply described this way: in a universe that is infinitely big and exists forever, sooner or later histories will repeat, and so there will inevitably be other copies of each of us hanging out there. Vilenkin's calculation with his colleague Jaume Garriga, at the University of Barcelona, in Spain, almost twenty years after he discovered eternal inflation, was more substantial, precise, and subtle than that, however.

According to quantum mechanics, in any finite region there can only be a finite number of distinct quantum states. This is because at the subatomic level, energy comes packaged in discrete units—the quanta of quantum theory—as we learned in Chapter 2. An infinite continuum of energies is simply not allowed. That means that within a finite spacetime region there can only be a finite number of distinct histories, or a finite number of past events that can be distinguished from each other, even in principle. In 2001, Vilenkin and Garriga estimated the number of possible histories that could have existed since the big bang in a region the size of ours to be something around 10^{150}. "It is a crazy number, but the important thing is that it is finite," says Vilenkin.[8]

Meanwhile, eternal inflation generates a limitless number of regions just like ours. "So, given a finite number of histories playing out in an infinite number of regions, histories must repeat," says Vilenkin, "precisely."

This is not the same as the parallel universes that might occur according to the many-worlds interpretation of quantum mechanics, which we encountered in Chapter 2. There, in some sense, our actions cause reality to split, creating different versions of ourselves branching out into parallel realities, where different choices are played out in each. Our parallel selves in the inflationary multiverse are more like accidental replicas, and because there is only a finite number of possible fates to be handed out, those replicas are doomed to relive our lives in almost the same ways—or maybe even exactly the same ways, just in another time and place.

Even then, there is ambiguity about what it means to say that those lives happen elsewhere, in another time. Vilenkin thinks back to lessons from Einstein's theory of relativity, which we encountered in Chapter 1. According to relativity, we cannot define a distinct "now" that encompasses all observers and all places in the universe. There is simply no unique ordering of events that every observer must agree on and no objective concept of simultaneity. There is no universal clock that governs the multiverse.

This ambiguity about time becomes even more pronounced when you think about two different bubble universes that are being dragged apart at faster-than-light speeds by an inflating background. It is impossible to send light signals between them to communicate, so it would be impossible to even imagine synchronizing our watches. Therefore, Vilenkin argues, it is meaningless to try to argue that the events that occur in another universe take place at a different time; we may as well think of our clones playing out their lives right now, alongside us.

It's weird to think that there are other mes out there, living the same life as me at this moment, but it does not trouble me deeply, as I feel no connection to these doppelgängers. Their existence just seems coincidental. (By contrast, with the many-worlds interpretation of quantum theory, I do feel a responsibility to any quantum clones I make, because my choices brought them into being.) Yet this notion clearly chilled Vilenkin. Why?

"It actually made me upset," says Vilenkin. "We are going into philosophy, or even beyond philosophy—it is emotion." Why did he feel the impact of this so viscerally? "I found it upsetting that we lost our uniqueness," he explains softly. On the cosmic scale, we are nothing more than one more copy on a factory conveyor belt. This realization cut to the core of Vilenkin's sense of self and of how he might derive meaning in his existence. "We do not have any significance—this particular copy of the Earth, on the cosmic scale," says Vilenkin. Instead, if we want to find meaning, "to explain why we are and to find our reasons to be, we should look for them locally," he says.

Again I remain unshaken. Perhaps the difference is that I subscribe to the notion that we have free will and so my identity is not dictated by the accidental amalgam of particle motions and collisions, set out by the big bang, that created a chain reaction leading to inflation, stars, galaxies, planets, people, and me. So even if there are other inflationary copies of my physical self out there, they cannot be *me*. (We'll come back to this discussion about free will in models of physics where there are multiple versions of ourselves in Chapter 9.)

Vilenkin has since also made "a little more peace" with this revelation after he was contacted out of the blue by a biochemist who studies the origin of life and had just read of Vilenkin's work. He had been pondering how the circumstances for life to spring forth may be exceptionally rare, but he wrote to Vilenkin to explain that if the multiverse view is correct, then he was comforted to think that life will inevitably arise someplace. This short communication gave Vilenkin back a sense of purpose for humanity and placed into perspective our relationship to the pocket of the multiverse we call home. "If indeed the origin of life needs such a rare fluctuation that it has indeed happened only once in our observable universe, then we have quite a responsibility, because if we screw up, that's it," says Vilenkin. "So we are responsible for a large piece of real estate, and we can either vanish or maybe colonize a good part of the galaxy or beyond."

Vilenkin was still thinking on a small scale, though. What about the option of creating your own bubble universe and potentially colonizing that, communicating with it, or mining it for resources? These were ideas that would start to fascinate Linde, as we'll come to in Chapter 9. But first a baby universe has to be made. Vilenkin had shown how it may be possible to create a universe from nothing. Might it not then be even easier to create a baby universe from something here on Earth? Guth and his students were about to answer that question.

6

The Accidental Universe Makers

When you first set eyes on the ATLAS detector, the largest experiment housed along the 27-kilometer tunnel of the Large Hadron Collider, you are immediately struck by its vast size and its immense beauty. On my first visit to the particle accelerator, which is run by the CERN laboratory, located on the border between Switzerland and France, back in 2009, Sergio Bertolucci, then CERN's research director, urged me to visit the LHC's subterranean heart before it was closed off for data taking. "It's magnificent," he said, "like standing within a cathedral."

ATLAS is one of four experiments dotted around the LHC's circular track. The apparatus stands 25 meters high, about the size of a tall apartment building, and weighs 7,000 tons. Its octagonal symmetry and the vivid colors of the wires, magnet casings, and superconducting devices that form the detection chamber call to my mind Middle Eastern mosques, which are decorated with bright geometric tessellations because Islam forbids icons of people or God. I quietly wonder what archeologists excavating the site far in the future will ascribe as the purpose of the LHC: science or sanctuary?

The LHC has been dubbed the "big bang machine"—but it was not designed to make the baby universes described in this book. Instead, it earned the nickname because it seeks to probe the physics that takes place at high cosmic energies soon after the big bang, but not at our universe's actual birth. The fact that according to some physicists it may one day be possible to actually make a universe in a particle accelerator is a happy coincidence.

To rev particles up to these colossal energies, the LHC accelerates them to close to the speed of light and then injects them in two beams that travel in opposite directions around its circular track. These beams cross at four points on the circuit; the ATLAS detector is situated at one of those meeting points, and three other experiments (CMS, ALICE, and LHCb) stand at the others, each hunting for a cornucopia of exotic new entities. It's in these collision zones that energies rage high enough to generate particles and objects that were seen in our infant cosmos.

In 2012, to worldwide fanfare, LHC physicists announced that they had succeeded in creating the Higgs boson, the one remaining puzzle piece in the standard model of particle physics. That theoretical framework had been more or less sketched out by the end of the 1970s, and encompasses three families of particles: the leptons, including electrons and neutrinos; the quarks, which bind together to form protons and neutrons in atomic nuclei and can also come together to form other, less familiar particles; and the bosons, including the particle of light, the photon.

The bosons can act as mediators between other particles, and by so doing, they transmit force. On a microscopic level, physicists picture the electromagnetic force between two electrically-charged particles as being carried by the constant exchange of photons between the pair, either bringing them together or pushing them apart. Similarly, the strong force that sticks quarks together is mediated by gluons (named for their adhesive properties), while the W and Z bosons are associated with the weak force that controls radioactive decay.

By 1995, various particle accelerators around the world, running at lower energies than the LHC, had found all predicted particles except the Higgs boson, which is associated with the Higgs field that gives particles their identities (see Chapter 3); seventeen years later, LHC physicists sifting through the debris of their collisions were able to confirm that both the ATLAS experiment and its rival CMS experiment, also at the LHC, had generated the Higgs boson, which flirted with existence briefly and then decayed into more common daughter particles in the detection chamber.

Now that the standard model's zoo of particles is complete, it might seem like the LHC's work is done. But the machine powers on; in 2015 it began a new run, smashing particles together at 13 teraelectron volts (TeV), which almost doubles the energy used to produce the Higgs boson. One reason it still has work to do is that it is searching for evidence of theories that go beyond the standard model. As we'll see in more detail in

Chapter 7, which talks about one such popular extension, string theory, there are good reasons to believe that the standard model is incomplete. The LHC is hunting for evidence that could support or rule out any number of candidate models.

Various theoretical extensions to the standard model predict a wealth of oddball particles and bizarre entities that have never been seen, including the monopoles that Alan Guth was chasing in Chapter 3 and mini black holes. These are of particular relevance for our purposes because, as I am about to discuss in this chapter, a new universe created at a particle accelerator could be mistaken by us, from the outside, for a mini black hole. As we will also see in Chapter 8, a monopole might act as a seed from which to grow such a new universe. LHC physicists are currently on the lookout for any such strange objects that may appear fleetingly in their machine.

I have returned to the LHC to meet with Eduardo Guendelman, who is one of the architects who designed the blueprints for baby universe production—and who is also one of the project's strongest adherents. Usually based at Ben Gurion University in Beer Sheva, Israel, Guendelman is attending a particle physics conference at CERN, and he's invited me along on the official conference excursion to the village of Chamonix, on Mont Blanc in the Swiss Alps, about 1,000 meters above sea level. I joke that in terms of the analogies for inflation I have been using in the book, a village perched up the side of the mountain corresponds to the cosmos being in a false vacuum—the precarious high-energy state that caused our early universe to inflate. Let's hope it doesn't come tumbling down.

As we wait in a CERN parking lot for the bus to take us partway up the mountain, along with about a hundred other particle physicists, Guendelman stands out from the crowd. He is a tall man, head and shoulders above most others in the group, with a long beard of curling brown hair and a cap covering a balding head. It's the height of summer in Europe, and today is warm, 30 degrees Celsius (86 degrees Fahrenheit) at sea level—but Guendelman is huddled in a warm jacket and shivering. He tells me that in Beer Sheva the temperature was 45 degrees C (114 degrees F) before he left, so it's understandable that he finds this weather chilly. The LHC is housed under the ground we're standing on, smashing particles as we speak—and it is in just such a particle accelerator that a baby universe might be made. So I ask Guendelman how it feels, as a budding universe maker, to think that a cosmos could be being born below our feet at this very moment. "Weird," he replies succinctly. Guendelman is a taciturn man.

Unlike those I have interviewed in earlier chapters, Guendelman is not as used to the limelight. So while others I have spoken to are well versed in translating their work into digestible soundbites for the public, Guendelman has no such training in spin. Instead, he talks almost entirely in mathematics, backing up each statement with a set of equations and diagrams. He has brought a selection of relevant papers for us to go through on the bus, whereas I've only brought a warmer coat, gloves, and wooly scarf to layer on as we climb the mountain, and my voice recorder. I need to learn an entire new coordinate system during our bus trip to keep up with him. The fact that Guendelman insists on rigor is refreshing. Since the topic of our discussion is how to make a small universe under our feet—and how to be sure it won't expand so much that it swallows Geneva, Europe, Earth, the galaxy, and the rest of our universe in the process—he needs to be able to back up his arguments.

In the bus, Guendelman recounts his childhood in Santiago, Chile. His father was a tailor, while his mother loved painting. There was little doubt he would become a scientist, though, because his two older brothers were studying medicine and engineering, and encouraged him in that direction. Unlike them, however, he chose a more abstract and less vocational direction, inspired by the push toward astronomy in his hometown. Chile, with its temperate climate and lofty mountains, has become an ideal location for establishing world-famous telescopes, and so even as a boy, Guendelman knew that astronomy had a real future.

Guendelman joined an astronomy club, where he quickly realized that he was perhaps better suited to theoretical physics. He recalls a reluctance to polish the lenses of telescopes, in case he damaged them, and indeed, in later years as an undergraduate at the University of Chile, in Santiago, he had a habit of breaking lab equipment, "though nobody died," he laughs. For graduate school, in 1980, he moved to MIT, which offered far more opportunities physics-wise than he could find in Chile, but was vastly more competitive than his undergraduate university.

Guth arrived at MIT as a young professor just as Guendelman was preparing for his general exams. He had started to make a name for himself as the author of inflation theory but had not quite reached superstar status; that would take another few years of Guth giving multiple talks each week in the Boston area, publicizing his theory.

Looking back now, it seems like a no-brainer that a young graduate student should want to align himself with the assistant professor who had a hot new cosmological theory. But inflation had not actually grabbed Guendelman's imagination at that point. To him, it was just another idea—albeit a good one—in a sea of possible theories about the early universe, which may or may not one day be proved right. But Guendelman did find the novel theory interesting enough to merit further investigation for his PhD dissertation. More important, when it came to picking an advisor, "Alan was friendly," he remarks. "That's not exactly a scientific reason for choosing an advisor, but you have to work with somebody that is going to treat you well." So Guendelman decided to study with the amiable Guth, without realizing that his new mentor was about to set him off on a quest to make a baby universe.

By now, the physics excursion has reached the Chamonix valley. As Guendelman and I, and about fifty other particle physicists, squash into a single cable car to take us up to the mountain station, Aiguille du Midi, at an altitude of almost 4,000 meters on Mont Blanc, he talks me through the project that Guth assigned to him, along with another student, Steven Blau, who is now the editor of *Physics Today* magazine. This is the problem that would eventually, and inadvertently, lead them to the realization that humans might actually make a universe in the lab. As the cable car lurches forward, taking my stomach with it, I'm pleased about the distraction.

In Chapter 5 we encountered Alex Vilenkin's work on how to create a universe from (almost) nothing—that is, from no space, no time, and no matter. (I include the "almost" because you do need the immaterial laws of physics to reign over your kingdom of material emptiness.) Guth was interested in what seemed by comparison—at least at first glance—to be a slightly less ambitious project: how to make a universe out of a tiny amount of something.

Inflation, Guth noted, could blow something very small up to, literally, astronomical proportions. It was this discovery that led him to quip that the universe is the "ultimate free lunch." In theory, you need only one ingredient to make a universe that would eventually populate itself with matter: a false-vacuum bubble. But just how small could that initial seed be? Putting aside the "engineering problem," as Guth refers to it, of where you would find that false-vacuum bubble, he felt it was important to clarify whether a truly tiny bubble of false vacuum really could turn into an adult universe, in order to iron out the mechanics of just how inflation might have begun in our early universe.

Recapping from Chapters 3 and 4, inflation theory posits that the universe is pervaded by an inflaton field—a cousin of the Higgs field. This field could contain any amount of energy, and could gain or lose energy over the course of cosmic history. Physicists visualize the path of the field, as it shifts energies, as if it were climbing up and down a series of mountain peaks in the Alps.

Today our universe is in the true vacuum—that is, the inflaton field is in its lowest energy state. They are, to continue the mountain analogy, at sea level. As we saw in Chapter 3, Guth's idea (and, from Chapters 4 and 5, Andrei Linde and Vilenkin's refinements) required that at some point in the early universe (or the multiverse) the inflaton field became temporarily lodged in a higher energy state. It could balance in this position for a while—like our temporary visit to Aiguille du Midi—but eventually it would come down to the true vacuum, or at least patches of it would, if you followed Vilenkin and Linde's logic that led to their development of eternal inflation. Our cosmos now exists in one such inflated patch of true vacuum.

It's worth noting that Guth was thinking about a situation subtly different from the one that Vilenkin and Linde hypothesized gave rise to eternal inflation, as laid out in Chapter 5. According to eternal inflation, there are many parallel universes—many bubbles of true vacuum—being flung apart in a false-vacuum multiverse. Those bubbles, according to eternal inflation, can and will occur naturally. But what Guth wanted to know was, if we start out in a universe filled with true vacuum, just like the true vacuum in our universe now, could we take a smidgeon of false vacuum—a portion of space primed to expand at faster-than-light speed—and watch it (or make it) inflate into a new universe?[1] It was that question that needed to be answered in the affirmative if physicists were ever to have any hope of making a baby universe in a laboratory.

The more anxious among you might be worried about the obvious health and safety issue that arises when you talk about inflating a new baby universe *within* a preexisting stable universe filled with true vacuum. If a bubble of false vacuum suddenly did appear in our universe—either naturally or by human design (such as in the caverns housing the LHC)—would it inflate and destroy our spacetime like a runaway cosmic airbag suffocating those riding in the car, and then potentially exploding out of the vehicle and expanding ever onward? That was an issue that needed to be addressed too.

It's not such a fanciful concern either. In 2014, particle theorist Joseph Lykken at Fermilab, in Batavia, Illinois, announced the results of some preliminary calculations he and his colleagues had carried out based on the information we now know about the Higgs field, thanks to the discovery of the Higgs boson. Lykken plugged the known mass of the newly found Higgs boson, 125 gigaelectron volts (GeV), into an equation that characterizes the Higgs field, which pervades our universe. He found that the field is on the cusp of stability. If the mass had been a fraction higher, then our universe would be stable, but as it stands, Lykken's team says, our cosmos is in a metastable state. That means that the Higgs field—just like the inflaton field in its false vacuum state—is precariously poised in a higher-energy vacuum and searching for a true vacuum to fall into. It would take just a quantum fluctuation in some distant galaxy to knock a patch of the universe into that stable vacuum. This would have catastrophic consequences for any living beings in the part of space where the vacuum changed, as conditions in the new vacuum are unlikely to be conducive to the life and galactic structure that evolved in the old vacuum. These would be snuffed out, like a candle flame suddenly starved of oxygen. And as Lykken told me when he announced his finding in a podcast interview for FQXi, once the transition to a new, more stable vacuum happens in a small patch of space anywhere in the cosmos, "it will then be energetically favorable for that bubble of the better vacuum, the low-energy vacuum, to keep expanding at the speed of light until it fills up the whole universe." In other words, a slight quantum nudge could cause our universe to switch vacuums in an instant, destroying us all in the process. Yikes!

Eternal inflation similarly brings with it some doomsday scenarios. Guth, Vilenkin, and others, including my boss at FQXi, cosmologist Anthony Aguirre of the University of California, Santa Cruz, have independently calculated that it may be possible for another bubble universe to be born right in our corner of the multiverse. If that happened, and if we're very lucky, it might leave an observable signature that would serve as the first evidence of the multiverse. Our rowdy neighboring bubble universe might gently bump our cosmos before it is swept away from us and carried deeper into the multiverse. Such a nudge could have left a scar on the cosmic microwave background, and indeed, cosmologists are scouring the skies for such signs. So far, there have been tantalizing hints—within the radiation pattern, for instance, researchers have found an inexplicably large region of lower temperature, dubbed the "cold spot," which could be a dent caused by a cosmic collision—but nothing that stands as definitive proof of the multiverse . . . yet.

If we're unlucky, then such a titanic crash could be more catastrophic; the violence of the smash could destroy one or both colliding universes. I interviewed Vilenkin about the possibility some years ago, for *New Scientist*, when it was first mooted by Aguirre, and asked him if death by bubble collision is something we should be worrying about. In his trademark deadpan style, Vilenkin responded that since death would be instantaneous, with no warning for the inhabitants of our cosmos at all, there really isn't any point losing sleep over it.[2]

It may not be worth getting anxious about cosmic apocalypses we have no control over. They are far less likely to strike than a devastating natural weather disaster. But this book is about intentionally creating a universe in a particle accelerator. The LHC faced enough bad publicity in 2008 when a lawsuit was filed in a US court to prevent its operation, out of fear that any mini black holes it may produce might destroy the planet. The case was dismissed because such tiny black holes are too small and would evaporate too quickly to present any danger.

But unlike a black hole (mini or cosmic), an inflating baby universe created from a patch of false vacuum is designed to get bigger, rapidly. If one was made—either intentionally by physicists in a lab or by chance among the random collisions at the LHC or a future, more powerful accelerator—would it prove lethal? I shall reveal now, so as not to cause unnecessary distress to readers of a more nervous disposition, that Guendelman's calculations indicate that baby universes pose no such threat.[3] But why not?

We're now almost 4,000 meters above sea level, stepping out onto a viewing platform on the frosty mountainside at Aiguille du Midi. The air is icy, as you might expect at such a high altitude. I say "might" because, as it turns out, I am one of only a few in our party of mostly theoretical particle physicists who have considered that this journey might entail a sharp temperature drop. As I layer on the coat I carried with me, wrap myself in a scarf, and put on a pair of gloves, one of our group, wearing sandals, a T-shirt, and shorts, remarks: "You look so well prepared. How did you know it would be cold?" I'm not sure I want to explain to a physics professor that at higher altitudes the atmosphere is less dense, so air pressure decreases, along with temperature; I figure it's something a physicist should know. So I just point out that I noticed the mountain was capped in snow.

Pressure changes are also key to a paradox about the fate of a false-vacuum bubble that Guth set Guendelman and Blau to investigate. Guth wanted to know how small such a bubble could be to seed a new universe. But when you took the pressure exerted by the false vacuum into account, along with the energy inside the bubble, it appeared that the bubble should both inflate into a full-blown adult cosmos and collapse into a black hole at the same time.

To see why, remember that a bubble of false vacuum has a constant positive energy density, which should drive it to inflate. Recall from Chapter 3 that when a normal balloon filled with helium, say, increases in size, the helium particles have a bigger volume to explore, so the density inside the balloon drops. Not so in the case of a balloon filled with an inflaton field that is in a state of false vacuum, however. In that case, as the balloon increases in size, the field replenishes the energy inside it, in order to keep the energy density at a constant level. This replenishment creates a gravitational push outward, forcing the balloon to grow—hence inflation.

But false-vacuum bubbles have another peculiar property, which I skated over in Chapter 3, that in theory should have the opposite effect: the false-vacuum bubble exerts negative pressure on the bubble wall, trying to suck it inward again. To understand how, imagine holding your false-vacuum balloon in your hands and pulling its walls outward. Once again, as the volume inside the balloon gets bigger, more energy must be pumped into the balloon to keep the false-vacuum energy density inside it a constant. Where does that energy come from? The only place it can come from is from you, since you're the only other player in the game. You must be losing energy—and working up quite a sweat—as you pull the walls outward. If pulling the walls out is such a tough job, it's fair to reason that that's because the walls are fighting you and trying to contract at the same time that you are trying to make the balloon expand. That means the false vacuum is exerting a negative pressure, pulling inward.

During inflation, when our early universe was in the false-vacuum state, this negative pressure would not have been too troubling. If the false vacuum fills the entire universe (or multiverse), then this negative pressure will not cause the universe to contract, or even work against inflation by slowing down the expansion. Pressure comes into play only where there is a pressure difference between one side of a barrier and the other, with the barrier being pushed more from the side where there is higher pressure toward the side where there is lower pressure, until the pressures on both sides equalize. If there's nothing outside the false vacuum bubble—no space beyond the edge of our universe—then there's no pressure difference to worry about.

But if you are thinking about creating a false-vacuum bubble within an existing universe—especially one such as ours, which is filled with true vacuum—the pressure difference between the outside and the inside of the bubble would be crucial. The negative pressure inside the bubble would force the bubble walls to contract. A sensible answer is that the fate of the baby universe depends on the balance between the two effects: if the outward push is greater than the inner pull, the bubble would expand, but if it is the other way around, it would contract.

Indeed, when Guendelman, Guth, and Blau settled down to work out which effect would dominate, they found a tipping point. If the initial bubble is below some critical size and mass, 10^{-26} centimeters across and weighing just about 1 ounce, or around 25 grams, the false-vacuum bubble would immediately collapse down into a black hole. It would never inflate.

But what the math told them happened above that critical mass only served to confound them. Einstein's equations showed that a false-vacuum bubble that started with a mass above the critical value must paradoxically both expand and contract at the same time. "A false vacuum bubble should expand, but if it is surrounded by a region of higher pressure it should contract—it is a contradiction," says Guendelman. "So you have to solve this to make these two statements agree."

It did not help that the team could not visualize what this meant because the math they were using could only tell them that from the outside the bubble just looked like the solution for a black hole in Einstein's equations. What went on within was shrouded.

To see why they hit a dead end, it's worth taking some time to look at how black holes are formed in the universe normally. There are two types of large-scale black holes in our cosmos: stellar black holes weigh in with masses between a handful of Suns up to a few tens of times our Sun's mass, while supermassive black holes, which reside at the centers of some galaxies, including our own Milky Way, can be billions of times heavier.

Stellar-mass black holes form when some stars die. Stars are powered by nuclear reactions in their core, but when their fuel is spent, they begin their demise. In the throes of death, the star begins to burn fuel in the shell around its core as its surface explodes outward in a supernova explosion. Our own Sun will go through this death cycle in about five billion years, eventually shrinking down as gravity pulls its constituent matter back inward. Without the nuclear power that maintained its original size, it will form a much more compact stellar remnant known as a white dwarf, roughly the size of Earth (which by then will have been snuffed out by the Sun's own supernova explosion) but a million times more dense. The corpses of stars that were somewhat bigger than the Sun are denser still. These stars collapse to become neutron stars, which can pack a whopping three times the mass of the Sun into a sphere with a radius the size of a city, just around 100 kilometers.

Stars that are even more massive form black holes, as gravity crushes their matter into a singularity—an infinitely small and dense point. This singularity is so dense that it distorts the fabric of spacetime around itself in such a way that anything that passes nearby will be drawn inexorably toward it; any matter will be ripped into its constituent pieces and crunched down. Even light rays, the fastest movers in the universe, will be pulled in if they veer too close. The boundary of no return for light is called the event horizon, which can be pictured as an imaginary sphere encasing the black hole.[4]

The singularity at the black hole's heart cannot be described by Einstein's equations or any other known laws of physics. It's similar to the singularity that, as we saw in Chapter 1, Stephen Hawking, Roger Penrose, and George Ellis had argued back in the 1960s must lie at the origin of our universe. The only difference is that the singularity at the birth of our universe was a white hole—a black hole running backward, spewing out energy rather than devouring it.

Astronomers do not know exactly how a supermassive black hole is made. One possibility is that a large stellar black hole can grow to epic proportions after chowing down on other nearby black holes. We'll set aside supermassive black holes for now, but we will revisit them in Chapter 9, when talking about ways to detect a baby universe inside a black hole in the lab or to uncover a parallel universe hiding in a cosmic black hole in the sky. At any rate, like their smaller siblings, supermassive black holes are believed to have a lethal singularity at their core shrouded by an event horizon that marks the surface of this cosmic behemoth.

The event horizon was precisely the roadblock that Guendelman and the others hit now. They had a solution that worked mathematically, as mentioned above, which told them that the bubble wall should simultaneously move both outward and inward. But it made little sense to them when they tried to visualize what was happening because they appeared to be staring at the mathematics of a static black hole event horizon that was not moving in or out.

The team was stumped for months, painstakingly checking their calculations to find any possible source of error. Then they hit upon the idea of reexamining the problem from a different angle, one that could let them peer beyond the event horizon of a black hole, if not quite into its heart. This perspective shift involved changing the coordinate system that they were using to analyze the bubble, a trick physicists often use to help illuminate murky situations. There's nothing shady about this move; it's just that some ways of marking distance are more helpful than others depending on the geometry of the situation you find yourself in. For instance, if you were standing on the equator and a friend texted you with instructions to meet at her house, which she says lies 1 kilometer directly south from you, you'd be all set. But if you received the same instructions while standing at the North Pole, you'd have a conundrum, as "1 kilometer directly south" from that location would pick out a whole circle of latitude, not a single point on the globe.

In a similar way, Guendelman and company had found that there was some ambiguity about what happened to an object—a spaceship, say—that passed through the bubble's event horizon. They hoped that switching coordinate systems might reveal where it went. If the bubble universe was truly equivalent to a black hole, as their first calculations seemed to suggest, then—in the new coordinate system—they would be able to track the ship passing through the event horizon and then being drawn inexorably toward the singularity at the black hole's center. But that's not what the coordinate change revealed.

They examined the path of their spacecraft moving toward the bubble wall from an infinite distance away. At first, as expected, their calculations showed that the distance between the ship and the bubble decreased. If the bubble was just a black hole, then this distance should have continued down to zero, when the ship eventually crunched into the singularity.

Instead, what they saw was an effect that Guendelman calls the "doubling of space." At first the distance between the ship and the bubble decreased as the ship approached the wall. The distance continued to drop after the ship passed through the event horizon. But, crucially, it never got down to zero. The ship never hit a singularity. Instead, the distance reached a minimum value and then started to increase again, as though the ship was exploring a whole new vast space.

The team realized that in this particular instance, when examining the geometry of a baby universe, they were looking not at the geometry of a black hole at all but at the geometry of a wormhole.[5] These cosmic objects have become the stuff of science fiction legend, serving as shortcuts between vastly separated parts of spacetime or, in this scenario, as bridges between two different universes.

The term "wormhole" was coined by Princeton physicist John Wheeler in the 1960s. But these objects were originally discovered as solutions to Einstein's equations by Austrian physicist Ludwig Flamm a century ago, in 1916. Twenty years later, Einstein and physicist Nathan Rosen rediscovered a similar idea. To see how wormholes can arise in general relativity, imagine a flat spacetime containing two distant singularities—perhaps many millions of light-years apart. (Here I'm relying on a trick that physicists often use: it's too difficult to visualize how four-dimensional spacetime bends, since we're constrained to seeing only three dimensions, but you can think about how a two-dimensional cross section of spacetime might warp.)

General relativity allows for the possibility that spacetime can be folded, like a laptop. As spacetime folds, the two originally far-flung singularities can line up one above the other. The two singularities start to bleed toward each other, joining to form a wormhole—a shortcut that will allow you to pass from the top layer of spacetime to the bottom layer, without having to traverse the surface between (Figure 2). In this picture, spacetime exists only on the two-dimensional surface of the fabric. The void that lies between the folds is a no-go region.

This was the geometry that Guendelman, Guth, and Blau were seeing in front of their eyes. In movies and novels, the truly exciting function of

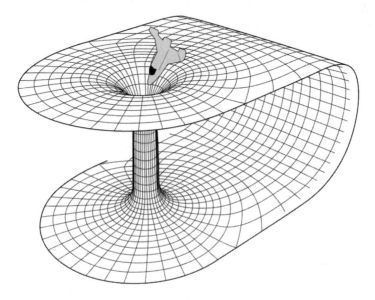

Figure 2: Wormholes act as shortcuts, connecting otherwise distant regions of spacetime.

The Birth of a Baby Universe

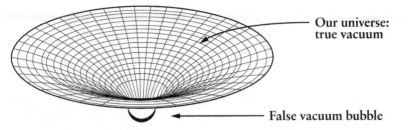

Our universe: true vacuum

False vacuum bubble

Figure 3a: A small bubble of false vacuum in our universe warps spacetime. The false vacuum inside the bubble pushes the circle to expand outward, while the difference between the pressure inside the bubble and outside it tries to make the circle contract.

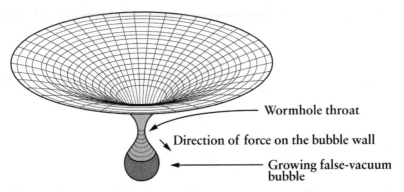

Wormhole throat

Direction of force on the bubble wall

Growing false-vacuum bubble

Figure 3b: As the false-vacuum bubble grows, spacetime distorts to create a wormhole throat. The spacetime geometry has flipped, so now both the false vacuum inside the circle and the pressure difference are pushing the circle to inflate. Although the bubble is increasing in size, it does not grow into our universe.

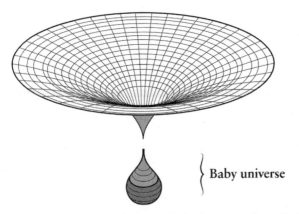

Baby universe

Figure 3c: Eventually the wormhole throat pinches off. The bubble of false vacuum has become a baby universe.

Credit: Above illustrations based on drawings by Eduardo Guendelman that appeared in Steven K. Blau, E. I. Guendelman, and Alan H. Guth, *Physical Review D* 35, 1747.

wormholes is that they let you travel quickly to distant realms. For Guendelman, however, the mind-blowing feature of the wormhole was that it turns spacetime inside out. It may not immediately be obvious how this happens, or why it's so important, but it was the key to unlocking the paradox of baby universe growth.

To see why, this time you need to imagine sticking a circle of false vacuum, rather than a singularity, onto a spacetime surface that is otherwise filled with true vacuum (Figure 3). (In three dimensions, the false vacuum bubble would be a sphere, but we're looking at the bubble's two-dimensional circular cross-section.)

From the outside, the false-vacuum bubble acts like a black hole, with the edges of the circle marking the event horizon. Just like a black hole singularity, the bubble warps spacetime, bending the flat fabric so that an indentation forms, protruding below (Figure 3a). At this point, the false vacuum inside the bubble is pushing it to expand outward, but the pressure difference is trying to force the edges of the circle to contract inward.

But then something cunning happens. As the bubble inflates, the indentation deepens, creating a wormhole throat—a thin tunnel that now links the small inflating bubble to the original parent universe (Figure 3b). The bulging bubble becomes a daughter universe in its own right, and its geometry flips. This is a crucial change because now the pressure difference pushing from the true vacuum to the false vacuum is actually pushing the bubble to expand, rather than contract. Both the internal force within the bubble and the pressure on the walls are now working in tandem with one joint aim—to inflate the bubble.

But, also thanks to the wormhole, the daughter's expansion will not encroach on the parent spacetime. Instead, the false vacuum bubble expands by creating its own spacetime that grows into the void—the absence of spacetime—between the universes. The wormhole is an umbilical thread joining parent and child, and it will eventually sever, disconnecting the inflating baby universe from its makers entirely (Figure 3c).

It was then that Guendelman was struck by the implications of what they had found. This question about building a universe had shifted from an abstract theoretical musing to something that could be carried out in a real-world laboratory. If you could get hold of a bubble of false vacuum with a mass of just 25 grams—never mind for a moment where you would find it—it would inflate into a whole new world in front of your eyes.[6] "This critical mass was a certain finite quantity, so you can therefore construct an infinite universe out of a finite amount of energy," says Guendelman. From the outside it may appear like a black hole. But within, it could be just as rich and varied as our own cosmos. "Once you get this result, it seems natural to say, oh, this provides a possibility of creating a universe in a laboratory," Guendelman continues. "Then why not think about making it right here?"

Guendelman may have wanted to shout the discovery from the rooftops, but Guth soon dampened his spirits. Guth may have been young and friendly, but he was a stickler for rigor and would not allow the team to go public with such a controversial claim without being incredibly sure it was right. "It was strange, in some ways, that such a crazy idea could come from someone so methodical," recalls Guendelman.

In hindsight, Guendelman appreciates his mentor's conscientiousness, but as a student he found it frustrating. "Alan doesn't like a good paper, he likes a perfect paper," says Guendelman. As far as Guendelman was concerned back then, their result was significant enough not only that they could write it up for an academic journal but also that he could complete his PhD thesis by 1984. In fact, the paper was delayed by Guth's repeated checks until 1987, when it was published in the journal *Physical Review D* under the muted title "Dynamics of False Vacuum Bubbles."[7] While the paper, with its somber title and technical abstract, did its best to underplay its hubristic conclusions, the idea has been at the forefront of Guendelman's mind ever since.

We're back in the bus now, waiting for it to leave Chamonix and return to CERN. It's ten o'clock at night, but we may be in for a long wait because the predesignated time for departure was set at eleven. Still, all but four of our party have returned to the coach, and the rest are getting impatient, arguing that they want to leave now, without the missing physicists. It seems a little harsh, and the atmosphere is getting pretty tense, because the bus driver—quite fairly, in my opinion—wants to wait for the others to return, and leave at the originally scheduled time. It's not actually the stragglers' fault that they are not here early.

As the arguments get ever more heated around us, I take the opportunity to ask Guendelman if he has ever paused to think about the ramifications of being a universe builder. If a cosmos we create becomes populated with intelligent life, shouldn't we worry about the potential suffering of those beings? "If we don't make the universe, they wouldn't have existed at all, so why should they complain?" he says with a smile.

Aware that this sounds flippant, Guendelman elaborates. He has children, and he appreciates that a very similar question is something that any parent has to think about before deciding to bring a new life into the world, where that child will inevitably encounter pain—sometimes at a horrifying level. And every child brought up within a religious household will one day turn to his or her parents and ask them why a benevolent God allows evil in the world. "In the old days we were told that God takes the view that learning requires pain, that God thinks, 'If that's what it takes, let it be,'" Guendelman says. "I'm not sure if that's right, but it is one view."

Guendelman often gives cosmology talks at his local synagogue, but says he tones down many of his more provocative ideas for fear of causing offense. But while he is reticent, he will say that he feels that some physicists go too far when they say that cosmology has proved that God cannot exist or that religion is antithetical to science. "You can turn the argument on its head and say that Einstein failed to discover the big bang, although it was all there in his equations, because he was prejudiced against the idea of the universe having a beginning because it seemed too religious," says Guendelman.

I am aware that I am putting Guendelman on the spot here and asking him to make pronouncements in an arena where his equations cannot help him. In Chapter 10, I'll return to the ethics of potentially producing life in a baby universe, highlighting the perspective of philosophers whose day job is to consider such issues.

Now it's 10:55 p.m., just five minutes until, we hope, the missing four people in our party will turn up at the meeting point, as they were instructed. But then an odd thing happens: having withstood fifty-five minutes of heckling from an increasingly rowdy busful of tired and demanding physicists who wanted to leave immediately, the driver capitulates and starts the engine, and we head back down the mountain to CERN. It seems bizarre that we waited almost an hour for the missing physicists to show up but then left five minutes before they were due to arrive. Sometimes physicists, famed for their logical minds, can be utterly irrational. Nobody actually knows who the absent ones are—no list of names was taken when we set out—and as the bus drives down the mountain road, I picture four confused physicists turning up at the meeting point and wondering why they have been forgotten.

This brings another issue to mind: should we be muscling into the territory of the gods at all? We could create life, perhaps even by accident, with possibly no way of ever reaching out to it, communicating with it, or helping its beings. Does that make us irresponsible deities, willing to just ignore our creatures the way we've just stranded those poor lost physicists in Chamonix, without even a note? "You could say the same about God creating and then abandoning us," says Guendelman soberly. "But now we are talking religion and we are getting into dangerous ground."

The next day I catch up with Guendelman in Geneva, to hear him continue his universe-making tale. As we'd left it, back in 1987 the prospect of creating a cosmos in the laboratory looked quite rosy. Yet, some thirty years later, we don't seem to have made a universe—or at least, not as far as we know.

That's because Guendelman's excitement had been somewhat premature. He, Blau, and Guth had succeeded in showing that a small bubble of false vacuum could inflate into an adult cosmos in the lab without displacing the surrounding space. But there was a major catch, which they picked up on thanks to Guth's insistence on delaying the paper: their baby universe fell afoul of a theorem first set out by Penrose in 1965.

Penrose stated his singularity theorem in terms of the formation of a black hole. It could be summed up as saying that if an object—a collapsing star, say—is contracting fast enough, it will inevitably crunch down into a singularity, the ultradense heart of the black hole. But if you reverse the argument, it says that if an object—say, a baby bubble—is rapidly expanding faster than some critical rate, then it must have originated from a white hole singularity.

Guendelman's math did show that to overcome its initial tendency to collapse, the baby bubble would need to be inflating at such a fast pace that Penrose's theorem would apply. This totally changed the complexion of the project. Somehow, it seemed, budding universe makers would have to get hold of, or make, a white hole singularity, to create a baby cosmos. That was not just going to be hard, it would be impossible. Penrose, Hawking, and Ellis had identified such a white hole singularity as our universe's $t = 0$: the big bang. By definition, there was no way to introduce a white hole and begin time within a universe in which the clock is already ticking.

It sounds like a monumental blow, but Guendelman shrugs it off as a mere inconvenience. "The singularity problem is what I call a second-order problem," he says. A first-order problem—a deal breaker for physicists—would have been if their calculations had shown that you would need to put infinite energy in to get an infinite universe out; there would be no way past such an energy block. Happily, though, the team had shown that, in theory, you only need finite energy to set off a universe. By contrast, the apparent requirement of a singularity was bothersome, but it did not shut down the project completely. It was just another hurdle on the road to success.

Since Guendelman is not a man who is prone to hyperbole or making wild claims, I feel I am able to trust that his optimism was justified. Indeed, within three years of Guendelman's stumbling on the singularity hurdle, physicists would find a way to bypass the singularity to give rise to a baby universe. In the next chapter we'll meet Joe Polchinski, one of the men who solved that problem by turning once again to a bit of quantum magic. But, having saved the baby universe project, he would then turn his back on it in favor of pursuing one of the extensions of the standard model of physics, string theory. Along the way, he and others would combine strings with eternal inflation to come up with a model some have argued kills the need for God entirely.

7

A Baby Universe, a String Multiverse, or God? Explaining Our Tailor-Made Universe

I n 1990, Joe Polchinski began to realize just how lucky he was—how blessed we all are, in fact. It wasn't so much the happy circumstances of his career as a young physics professor at the University of Texas at Austin that made him appreciative. It was more the chain of events that had led not just to his birth but to the existence of his parents, his grandparents, all his ancestors, the origin of life on Earth, the formation of the solar system, and even the production of elements in the universe.

The issue that seized him, along with many others, was why the universe seems to be perfectly designed to enable stars and galaxies to form in the universe—and ultimately for intelligent life to arise. The masses of elementary particles, the ratio of the strengths of fundamental forces, and the parameters that control the expansion rate of the universe all seem to be finely tuned to just the right values needed to evolve humans. As yet, no single "theory of everything" has been discovered that can explain why these variables take the convenient values they do. But if these parameters were slightly tweaked, our universe would be a very different and quite inhospitable place. The odds against everything happening the way it did by chance appear monumental, but somehow we all seem to have been winners in the cosmic lottery.

Could it be that the universe was made just for us? After all, it seems infinitely more likely that a universe that popped into being purely by chance would be entirely devoid of life. In the absence of any other expla-nation, for some, the fact that conditions in the universe seem to be hand-ily adjusted for us to enjoy has been presented as a strong argument that God must exist as the creator of this bespoke cosmos. Not satisfied with this theological conclusion, however, Polchinski turned first to baby uni-verse creation for an escape route. Perhaps, he and his colleagues conjec-tured, an initially hostile (and far more likely) parent universe arose, but then a (less likely) habitable child universe, *our* universe, was born from within it. For that to be true, they needed to investigate the same cosmic production method that Eduardo Guendelman and his colleagues had dis-covered (see Chapter 6) and find a way around the singularity roadblock that they had slammed into. And that's just what they did—with a little quantum help.

I have come to meet Polchinski at the University of California at Santa Barbara, a corridor down from Tony Zee's office, to ask him about his escape from the singularity problem. But—unlike Guendelman—having solved one obstacle to making a baby universe in the laboratory, Polchinski soon lost interest in the enterprise. In this sense, this chapter is a bit of a departure from the rest of the book: it explains why for many years the notion of making baby universes fell out of favor, even with physicists who had been briefly enraptured by it.

The reason for Polchinski's disenchantment with cosmic creation in the lab was simply that soon after embarking on it, he became embroiled in an alternative framework for physics that seemed to offer better answers to the fine-tuning issue that was frustrating him than solutions invoking baby universes (or, indeed, God) could. This rival model, the now much-vaunted string theory, posits that at the fundamental level, elementary particles aren't pointlike substances but really consist of tiny vibrating strings.[1]

Today string theory stands as one of the best candidates physicists have found for a theory of everything that combines gravity with quantum mechanics. Polchinski was one of the first physicists to tie strings to eternal inflation. As we learned in Chapter 5, inflation can spawn a multiverse of parallel universes—but these were long thought to be (physically) boring clones of the same parent universe, cooked up following the same recipe with little variety in their parameters, particles, forces, or even dimensions and laws. Polchinski and others realized that string theory adds spice, cre-ating a "string landscape" of unthinkably many alternative universes with a veritable cornucopia of different contents, dimensions, and laws.

For this reason, the string landscape has been touted as an alternative to the God hypothesis for explaining why our own universe seems weirdly suited for life. In the string landscape view, we no longer have to explain the existence of one lone, highly improbable cosmos with peculiarly perfect parameters for humans, perhaps forcing us to invoke the hand of God to create such convenient conditions. Rather, Polchinski argues, there are nigh on infinitely many different stringy universes, with myriads of different variables. Most cosmoses in the landscape won't be able to support life, but every so often one will pop up—such as our own—that can host people. That's because, if every possible type of universe can exist somewhere, then it should not be quite so surprising that one with our properties is among them. I want to ask Polchinski about this debate too: does string theory, coupled with inflation, really do away with the need for God?

The arguments coming from string theory were convincing enough for Polchinski to drop the baby universe project and focus all his attentions on this rival solution to the fine-tuning question. But ironically, as we shall see, it later turned out that if string theory is correct, it may make it far simpler to create a baby universe in the lab than physicists had originally thought—bringing this act of cosmic creation within human grasp.

It's hard to see the unrelentingly cheery Polchinski as one of the people who has been accused of killing God, albeit accidentally. In his early sixties, he is a keen cyclist, with a wiry athletic frame, a full head of gray hair, and a perpetual smile. He's filled with energy, fidgeting and swiveling in his office chair as we speak.

Polchinski has no particular beef with God. He describes himself as "the ultimate agnostic," explaining that he never makes assumptions without evidence. He was born in White Plains, New York, where his father was a Catholic stockbroker and accountant with an amateur interest in science but never the desire to pursue it in depth. His mother was raised a Protestant and, as was typical of the time, was a housewife, so she never got the chance to explore science. Both parents were "personally religious" without being churchgoing, and chose not to raise him within a Christian framework. "I was totally unexposed to religion," says Polchinski. "It's just alien to me."

Much more enticing than religion to young Polchinski was the How and Why Wonder Books science series, which he discovered when he was around six years old. He'd asked his parents for some for Christmas, and then, just before the big day, had been caught misbehaving. He cannot now recall what he had done wrong, but he still remembers feeling devastated when he was banned from using his presents for a few days as a punishment. Polchinski snuck in and read the books anyway.

At school Polchinski was good at math and science, but it was only as an undergraduate at the California Institute of Technology, in Pasadena, that he realized that theoretical physics combines both, allowing him to use math to unravel nature's mysteries. But by the time he hit grad school, in the late 1970s, it seemed as though most of the outstanding questions in physics had already been answered and it would just be a matter of years before the few loose ends would be tied up in a neat bow. The standard model of particle physics was taking form, as we saw in Chapter 6. The electromagnetic and weak forces had been unified and the workings of the strong nuclear force that binds quarks together to form neutrons and protons had been discovered. The Large Hadron Collider had yet to be built, but around the world smaller particle accelerators were finding particles that fit within this framework, verifying it again and again.

Although it would take until 2012 for particle physicists at the LHC to finally catch a glimpse of the elusive Higgs boson, completing the standard model's set, by the early 1980s everything seemed to be on track, with no surprise particles threatening to crash the party. "There'd been repeated confirmation of things we already knew," says Polchinski. So his years as a graduate student were spent learning about how quantum fields pervade the universe, and other ideas that formed the backbone of conventional physics.

Yet this continuing resounding success of the standard model was also a failure for physicists. Well before the 1980s, they'd realized that the standard model could not be the final word in physics. It provided no explanation for why the particles and forces have such convenient masses and strengths—just right to allow life to exist.

Take the strength of gravity, which played a vital role in drawing matter together in the early universe to make stars. If gravity were stronger, stars would burn their fuel more quickly and die earlier. Cosmologist Paul Davies at Arizona State University, in Tempe, has stated that doubling gravity's pull would make our sun shine a hundred times more brightly. This, in turn would shorten the sun's projected lifetime from roughly 10 billion years—it's currently about halfway through this span—to less than 100 million years. Davies has argued that this reduced life expectancy is too short a time to enable intelligent observers to evolve.[2]

Even at its current value, gravity still seems to be an immense force, holding us to the ground as Earth whisks us around the Sun each year. Yet a pin dropped on the floor can be lifted with just the aid of a child's magnet—the electromagnetic force is so strong that even a toy can outstrip the downward pull of the entire mass of our planet trying to keep the pin grounded. In fact, electromagnetism is more than 10^{30} times stronger than gravity. When the strengths of the four fundamental forces are compared, gravity is by far the puniest—and it's a good thing too, or else atoms as we know them would never survive against its efforts to squash them all together, which would prevent them from clubbing together to form coherent things like, you know, us.

I could go on and on listing more parameters that have been set to the ideal values for life to exist. But at the end, you could just shrug and say, "So what?" Worrying about why the universe's characteristics are one way and not another could be dismissed as an aesthetic concern, rather than a fundamental problem with our understanding of physics. But there were (and still are) other pressing puzzles that tell physicists that the standard model of particle physics definitely cannot tell us everything about the contents of the universe. Most notable of these are the discoveries of two mysterious entities over the past century, dark matter and dark energy—both of which we appear to observe, and yet neither of which seem to have any place in the standard model.[3]

Astronomers have known for decades that much of the matter in the universe is hiding out in some invisible form. This dark matter is confounding stuff that cannot be seen directly through telescopes, but its existence is apparent because it exerts a gravitational tug on nearby visible matter. Dark matter first made its presence felt in the 1930s, when astronomers noticed that stars orbiting the outer regions of some galaxies were whizzing around much faster than expected. Just as you can make a yo-yo swing around your head more quickly by tugging harder on its string—that is, by exerting a larger force—the best explanation for the outer stars' speediness was that they were being gripped by a greater amount of gravity emanating from within the galaxy than could be accounted for by adding up all the visible matter that astronomers could see through their telescopes. In 1933, the Swiss astronomer Fritz Zwicky attributed the source of this extra gravity to *dunkle Materie*, or dark matter. As far as physicists can tell, this unknown substance cannot be made from particles available in the standard model, so the standard model must be incomplete.

Beyond dark matter, evidence that physicists had misplaced a large chunk of the contents of the universe "blew up massively" in 1998, recalls Polchinski with a laugh. That's when two teams of astronomers, one led by Saul Perlmutter, at the Lawrence Berkeley National Laboratory in California, and the other by Brian Schmidt, at the Australian National University, in Canberra, discovered that the expansion rate of the universe is accelerating.

Their finding was one of the biggest shocks in physics because, based on Einstein's equations alone, there was no explanation for what might be causing this speed-up. Both teams came to their startling conclusion after monitoring certain types of supernovae that are created when stars that have a mass similar to our Sun's exhaust their fuel and detonate like a thermonuclear bomb. These type 1A supernovae are considered to be "standard candles" because they all burn with roughly the same brightness. That means that by measuring how bright or dim their explosions appear on Earth, astronomers can make a good guess at how distant they are, and thus how far their light has traveled to reach us and how long it took. Their emitted light waves are also stretched by the expansion of space while they make their journey to Earth, shifting their wavelengths to the red end of the light spectrum. This redshift gives astronomers a handle for measuring the expansion of the universe since the time that the light was released.

Both teams expected to find that the rate of the universe's growth was decreasing. Their thinking was that the universe would have undergone a brief period of rapid faster-than-light inflation soon after its birth in the big bang. Once inflation ended, the cosmos would still have been expanding, but less ferociously. Eventually the inward tug of gravity trying to pull the universe's contents back together should have taken over, gradually slowing expansion. So, the teams reasoned, measurements made today should show cosmic growth to be losing pace. Instead, however, both teams measured that the universe was expanding at an ever-increasing rate. In the absence of any definitive explanation of what was causing this, the acceleration has been attributed to an unknown "dark energy" pushing space apart; its identity remains to be revealed.[4] Astronomers currently estimate that 68 percent of the universe is made up of this dark energy, 27 percent is dark matter, and only 5 percent is ordinary visible matter.

It's actually the accelerated expansion of the universe that Polchinski was thinking about in 1990 that indirectly led him to look at baby universe production—although at the time he didn't know the phrase "dark energy" and the effect had yet to be measured. Polchinski is not psychic, nor did he personally predict the presence of dark energy in advance based on his own physics calculations. Rather, Polchinski credits his insight about dark energy to a colleague of his at Texas, Steven Weinberg, a physicist who had won a share of the Nobel Prize in Physics in 1979 for his role in uniting the theories behind electromagnetism and the weak nuclear force. "In all the time I've been in the field, the acceleration of the expansion of the universe is the most important discovery that's been made—and it was a surprise to everyone except Steven Weinberg," notes Polchinski.

How had Weinberg seen this effect coming? Back in the 1980s, before dark energy was observed to be driving the outward expansion of space ever faster, Weinberg had been pondering a physical quantity, called the cosmological constant, that particle physicists were sure must exist and would be pushing space apart at every point—basically doing exactly the sort of thing dark energy is thought to do.[5] The cosmological constant represents an intrinsic energy of space and was predicted to arise from quantum effects. Its origin lies in quantum uncertainty, which prevents us from ever being certain that spacetime is truly empty. The quantum vacuum, as we saw in Chapter 2, is actually constantly churning up pairs of virtual particles, which exist momentarily and then duck back down into the emptiness, disappearing forever. When particle physicists added up just how much energy these ethereal entities should contribute to the universe, they found that the cosmological constant—space's intrinsic energy—should be huge and should be forcing the expansion of the universe to accelerate at breakneck speed.[6]

On the other hand, Weinberg knew that at that time, in the 1980s, astronomers measuring the expansion rate of the universe had not spotted any such speeding up of cosmic expansion. We now know that's because their measurement techniques were not yet advanced enough. But at the time, this lack of evidence led astronomers to assume that the cosmological constant was most probably zero. (And even though we now have evidence that the expansion of the universe is accelerating, it's not doing so at anything like the pace predicted by the value of the cosmological constant calculated by particle physicists.) That meant that there was a huge discrepancy between the particle physicists' cosmological constant and the upper limit astronomers conceded it might take based on their measurements. The two figures disagreed by a whopping factor of 10^{120}. Again, physicists seemed to be getting something very, very wrong. They could not explain why the cosmological constant wasn't enormous.

Most physicists who engaged with the cosmological constant problem figured that the constant must be zero and that particle physicists who argued that it was gigantic were just doing their mathematics wrong. But Weinberg applied some lateral thinking to the problem to come up with a solution of an entirely different flavor. It was an answer that was (and remains) unpalatable to many physicists and has been denounced by some as little more than a nonscientific cop-out that bypasses a fundamental puzzle about nature by deeming it irrelevant.

Weinberg's argument started with the supposition that the final theory of nature, whatever it may be, does not actually prescribe one single value for the cosmological constant. Instead, he argued, it may only be able to provide a range of possible values that the constant might take, some as huge as the number predicted by particle physicists, and others approaching zero, to fit with the observations of astronomers. This spectrum of possible values might be realized in different parts of the universe or at different times, he supposed—though he had no way of explaining how or why that might happen.

Working with that assumption, Weinberg then argued that in a universe where the cosmological constant was large, cosmic expansion would have been so ferocious that matter would have been diluted out too quickly to have clumped together to form galaxies. The only value of the cosmological constant that could allow galaxies and structures to form would have to be almost zero, but not quite. Since humans obviously *do* exist and we can see galaxies around us, that must be the kind of universe we live in. Thus the cosmological constant must be small, Weinberg said, and he provided an upper limit to the size it would take.

And indeed, in 1998 when astronomers discovered that the expansion of the universe was speeding up, the amount of that acceleration they measured sat snugly with Weinberg's prediction of a small but finite cosmological constant. Today, although physicists still don't know the origin of the dark energy that is causing the expansion to speed up, a cosmological constant of the size predicted by Weinberg remains one of the best guesses as to its source. In this view, dark energy comes from the intrinsic energy of space.

In coming up with a value for the cosmological constant, Weinberg was invoking the anthropic principle, which is an argument that turns science on its head. Rather than using a fundamental model of nature to explain why the universe has the particular features that it does—or giving up and claiming that an unseen God waved a magic wand and made the universe just right for life—the anthropic principle starts from the fact that we know that humans are alive and well today and says that, as a result, we can be sure that our universe must have the parameters that are suited for life to exist. Perhaps, fans of the anthropic principle argue, there's no need to look for a deeper explanation than that.

This type of reasoning does not apply just to the cosmological constant; it can be used to wash away questions about the relative strengths of the forces when compared to each other, the masses of the elementary particles, and just about anything else that seems weirdly tailor-made for humans to evolve. Things are the way they are, anthropic reasoning goes, because if they weren't, you wouldn't be here to question them.

Polchinski's first reaction when he heard Weinberg's idea—before there was any observational evidence of dark energy to back it up—was to balk at the assertion. "This bothered me very much because we all hate the anthropic principle," says Polchinski. Physicists tend to strive to find fundamental explanations for why things are the way they are, including the values of physical parameters, such as the cosmological constant, which they hope are determined by some deep mathematical theory. "We all hate the idea that the constants of nature are random rather than predictable," Polchinski says. (Andrei Linde recalls that in the 1980s, the anthropic principle was so reviled that it was referred to as the "A-word.") But Polchinski was also uneasy because he had a niggling thought that perhaps there was an inkling of truth in this bold statement. "It was the first prediction I had seen that the cosmological constant wouldn't be zero, with an explanation for why it would be small," says Polchinski.

So with that in mind, in 1990 Polchinski began to search for a mechanism that could explain how our universe ended up with a tiny cosmological constant. It was this quest that inadvertently led him to dabble in baby universe building.

Polchinski turned first to quantum theory. Recall from Chapter 2 that physicists model quantum systems in terms of a wavefunction—a mathematical quantity that encapsulates the fact that before being observed, a system can take on a range of properties. An electron, for instance, can be in multiple places at the same time. It's only when you look at it that the electron's wavefunction collapses, snapping the electron into one location.

Perhaps, Polchinski thought, it might be possible to come up with a quantum wavefunction for the whole universe that similarly explained how the cosmological constant could have potentially taken any one of a number of values, ranging from zero to the ginormous value predicted by particle physicists. Like an electron settling into one location when it is observed, our cosmos might have randomly snapped into a state with a small value of the cosmological constant.

It turned out that Stephen Hawking had already had the same idea and had taken a stab at the calculations. The trouble was that when he followed his equations through, Hawking predicted that by far the most likely outcome in such a scenario would be a universe with no cosmological constant at all—no intrinsic energy to space, and so no accelerated cosmic expansion today.

Okay, that didn't help much. So then Polchinski's question shifted to how Hawking's universe—bereft of any cosmological constant—might morph into one with a small constant, of the sort that Weinberg had suggested would be the most nurturing for life. This is where Guendelman's work with Steven Blau and Alan Guth on baby universe production in the lab caught Polchinski's eye. As outlined in Chapter 6, those physicists had shown that it may be possible to make a whole new universe in the lab from a patch of false vacuum. Polchinski and his colleagues Dan Morgan and Willy Fischler, also physicists at the University of Texas, looked at their creation mechanism and began to wonder whether something similar might have happened spontaneously in the early universe. That is, could the universe have started out—as Hawking's calculations predicted—as a barren, cosmological-constant-free place, hostile to people, but then spontaneously have birthed a baby bubble with the cosmological constant that we need to survive? If so, perhaps we live inside that baby bubble.[7]

To be clear, Polchinski wasn't interested in making a baby universe in the lab. He just wanted to borrow the mechanism and the math that Guendelman and the others had set out, and use them to work out if we might live inside a baby universe ourselves. But in the course of his investigation, he and his colleagues found themselves staring at the same problem that had confounded Guendelman, Guth, and Blau: any bubble of false vacuum would fairly rapidly contract back down into a black hole unless it could be kick-started by a white hole singularity. Polchinski and his collaborators could no more inject a white hole singularity into Hawking's universe than Guendelman, Guth, and Blau could create one in the lab.

But Polchinski realized that there may be a way to sidestep the singularity issue. Once again it required a sprinkling of quantum magic. There seemed to be no way classically for a baby bubble to inflate into an adult universe unless it had originated from a white hole; it would collapse back down into a black hole before it reached adolescence. But there was a faint possibility that a baby universe could quantum-tunnel to adulthood without having to go through the pangs of puberty at all. That is, a small bubble might instantaneously become a large bubble without passing through a phase in which it's of medium size. After this tunneling, its adult form would be large enough to sustain inflation, turning it into a full-blown universe. Plus the quantum tunneling process would imbue this offspring universe with the energy needed to populate itself with matter and radiation, so it would truly be a cosmos that we could call home.

The team checked their idea mathematically by coming up with a wavefunction to describe the initial small bubble and working out the probability that it could make the quantum leap into a large bubble before classical physics forced it to collapse. The answer was that it was possible, but the probability was extraordinarily tiny.[8] "In some sense, that was okay, because it only has to happen once in the history of the universe," says Polchinski. All you had to do was wait. Back at MIT, Guth had also been struck by a similar quantum brainwave that might save their home-made universe project. Along with physicist Edward Farhi and a student, Jemal Guven, Guth too searched for a quantum tunneling solution to avoid the need to generate an initial white hole singularity in the lab, and they came to a conclusion similar to the one arrived at by Polchinski's team.[9]

The problem of how to get rid of the singularity and grow a baby universe was solved . . . kind of. Quantum tunneling could indeed get around the singularity problem, but there wasn't much chance of it happening spontaneously.

The take-home message was different for each team. For Guth, it showed that quantum mechanics could save the universe-building project, but not in a very practical way. The tunneling technique worked, but you might have to sit and wait around for billions of years for it to happen spontaneously, which did not bode well for the prospect of making a cosmos in a particle accelerator. That would require another ingenious step to catalyze the process, as we'll come to in Chapter 8.

For Polchinski, however, the calculation had served its purpose. He and his colleagues had used a baby universe to show how to get a welcoming, human-friendly habitat from a cold and empty one. The fact that you might have to wait billions and billions of years for the baby to blip into existence wasn't a big stumbling block, since the parent cosmos had all the time in the universe—quite literally—to give birth.

This is more or less where Polchinski parted company with baby universe building—and in that sense, I could just end this chapter here. But before we leap ahead to the next stage in the making of a child cosmos in Chapter 8, I want to stop and talk to Polchinski about the alternative model of reality that won him away from the project: string theory. There are two reasons I am interested in his work on strings. The first is that string theory, if correct, could mean it is much easier to make a baby universe at the energy scales that can be reached by a particle accelerator. That's because string theory suggests that tiny particles—and from our perspective, a baby universe made in the lab would count as a minuscule entity—experience a much larger force of gravity than we might expect from standard physics, and we need that gravity boost to help construct a baby universe. The second reason is that for Polchinski, string theory, combined with the inflationary multiverse, turns out to offer a better explanation for our finely tuned universe than his own earlier idea that we live in a baby universe bubble—or indeed the idea that God must exist and just made the universe that way.

To understand why string theory is so useful for universe making, we need to learn a bit more about its background. This was also one of Polchinski's aims as a young professor at Texas when he began reading up on the topic in tandem with his work on baby universes. Whereas his investigation into infant cosmoses was driven by a desire to explain fine-tuning, Polchinski was drawn to string theory because it promised to unite gravity with the three other fundamental forces of nature, electromagnetism and the strong and weak nuclear forces.

In Chapter 3 we saw how Guth began studying grand unified theories, speculative models that strive to bring together the forces described within the standard model—the electroweak interaction and the strong nuclear force—by positing that in the hot early universe they were merged into one single force. But even grand unified theories leave out gravity. Polchinski was looking for an ambitious framework that could bring gravity into the mix: a theory of everything.

Such a theory of everything, which combines quantum theory with gravity, is also needed if physicists are to ever figure out what happens in the singularities at the heart of black holes, or in the singularity at the origin of the universe. Within these points, matter is so dense and gravity so strong that Einstein's general relativity (or something similar to it) is needed to describe the warping of spacetime. But singularities are also unimaginably small, and so any understanding of what happens within them must also require an understanding of the quantum effects at play on microscopic scales. Neither theory alone can explain exactly what happens. They must be united.

Since quantum theory works so well for describing the behavior of particles on microscopic scales, and general relativity has proved a resounding success at explaining the origin of gravity and the motions of stars and planets on cosmic scales, it made sense to just stick the two sets of equations together.[10] But unfortunately, physicists quickly came unstuck when they tried the math. The problem was that when they used their combined equations to try to figure out what spacetime looked like on tiny scales, they got a bizarre answer: essentially, Einstein's spacetime fabric appeared to disintegrate into a roiling foam that fluctuated so violently it would tear itself apart.

You can get an idea of why they might get this strange answer just by looking at the equation for gravity that Isaac Newton developed in the seventeenth century, which showed that the strength of the force between two bodies is inversely proportional to the square of the distance between them. If you take two objects—Earth and the moon, say—and halve the distance between them, the force of gravity with which they tug on each other should increase by a factor of four. As you bring the bodies closer and closer together, the gravity between them rapidly shoots up. If that distance ever went to zero, then the size of the force would explode to infinity. This wouldn't happen in any circumstances that Newton was examining. But for quantum gravity equations that are trying to zoom in below microscopic scales, it's such a big problem that—according to the math—it would rip spacetime apart. Since spacetime, as far as we experience it every day, isn't such a turbulent mess, physicists knew something was going wrong in their thinking. "When you try to look at situations where both quantum mechanics and relativity apply, you find that taken together they are incompatible," says Polchinski. "They just give you nonsense that has nothing to do with what we observe."

The most promising candidate for a theory of everything that could get them out of this trouble, in Polchinski's eyes, was string theory. It has its roots in work by many people stretching as far back as the 1940s. But it really came to the fore in the mid-1980s as theorists began to realize that it could explain the properties of elementary particles as well as potentially uniting quantum theory with gravity.

String theory starts from a simple premise: all elementary particles are made from tiny vibrating loops. These closed strings define a minimum-length scale, close to 10^{-35} meters (less than a billion of a billionth of the diameter of an atomic nucleus), called the Planck scale. You simply cannot look below this scale. Shoving in this minimum length quieted down the equations of quantum gravity, so they no longer inevitably hit infinity when you try to see what happens on small scales. That's because you could now never move objects so close together that the distance between them went down to zero—which also meant that the gravitational force between them shouldn't skyrocket.

String theory was also appealing because instead of having to explain why there are a wealth of elementary particles—electrons, quarks, and so on, each with arbitrary masses—string theory says they all arise from one type of string. Their different properties manifest based on the different ways these strings can vibrate; particles are just different notes played by a single instrument. Polchinski was struck by the elegance of string theory. "It grabbed me from the beginning," he says. "All of physics comes from a single principle. Every other theory we've written down has a lot of knobs you can turn; they're not just given by principle."

All these benefits come with a price, however. The strings could mimic the behavior of elementary particles, but for the math to work out neatly, they needed a little extra wiggle room in which to vibrate. In fact, they needed six extra spatial dimensions to dance around in, meaning that if string theory is correct, then our universe must have ten dimensions in total: nine of space and one of time.

Those six extra dimensions exist throughout space, according to string theory, yet we experience only three spatial dimensions in everyday life because the others are too small for us to notice. Imagine looking at a carpeted floor from a distance; it seems perfectly flat, and indeed a person walking across the carpet would think of it as a two-dimensional surface. But a person who stopped, crouched down, and looked at it closely could see that it's woven from a number of threads. An ant traversing the carpet not only would be able to see those threads in magnificent detail but also would recognize that each fiber has both a length and a thickness. The fiber may even be thick enough so that the ant could walk around its circumference, accessing a hidden dimension that might easily have escaped our notice.

These extra dimensions could help explain why gravity seems so much weaker than the other three fundamental forces of nature. As I mentioned, electromagnetism is more than 10^{30} times stronger than gravity, and this bizarre imbalance between the forces is one of the finely tuned parameters that has enabled life to evolve. But if string theory is right, it could be that this relative feebleness is an illusion. In reality, gravity could be just as strong as the other three forces but only appears weak to us great big humans because it is leaking into these tiny hidden dimensions that we cannot access.

This is a crucial result for budding universe makers because it does more than solve the mystery of the weakness of gravity that we see around us. If extra dimensions are real, then elementary particles—like those ants walking on carpet threads—may be privy to them and, in turn, be able to access this leaked gravity. In Chapter 9 we'll see how that possibility led particle physicists to realize that such a gravity boost could make it possible to make mini black holes at the LHC. By the same token, it could also bring baby universe production within the reach of a particle accelerator.[11]

But by the time the twentieth century was drawing to a close, baby universe building was far from Polchinski's mind. After the announcement that dark energy was accelerating cosmic expansion, Polchinski—who by then had settled at UCSB—could have just sat back smugly and noted that his own work on baby universes, years earlier, provided a mechanism to explain how our universe could have gained this feature. Remember that the idea was (and still is) that dark energy likely has its roots in the intrinsic energy of space, pushing it apart—and that this intrinsic energy is quantified by a small cosmological constant. In Polchinski's baby bubble model, an originally barren universe, with no cosmological constant, would have burped out a baby cosmos—our universe—with a small cosmological constant of just the right size to enable the universe to harbor life.

Instead, however, Polchinski now turned to string theory with fresh eyes to see if it could provide a better explanation than a baby bubble universe for the value of the cosmological constant and for why our universe seems tailored for life. He was joined in this investigation by Raphael Bousso, a former student of Hawking's at Cambridge, who was visiting UCSB. Polchinski and Bousso recalled Hawking's attempt to come up with a theory that might allow for a spectrum of possible values of the cosmological constant, ranging from zero to huge. Could string theory supply an alternative mechanism that would provide that range?

The answer, again, came from string theory's six extra dimensions. Like zooming in on the fibers that make up a carpet, focusing in on any point in our everyday three-dimensional space should reveal these extra six dimensions (if we had a lens that was powerful enough). At each point in space, we would see the dimensions bundled together in any one of a vast number of compact six-dimensional shapes; they could be wrapped up to create the six-dimensional equivalent of a sphere, a doughnut shape, or a myriad of other six-dimensional pretzel configurations called Calabi-Yau manifolds (named in honor of American mathematician Eugenio Calabi, who first posited that such shapes might exist, and Chinese American mathematician Shing-Tung Yau, who proved that they could). The holes in the shapes are referred to as "handles," and each stable Calabi-Yau configuration generates a space with a different vacuum energy—a different value for the cosmological constant—depending on its shape, the number of handles, and other factors.

In 2000, Polchinski and Bousso together realized that if they considered Calabi-Yau manifolds with many handles, they could come up with a full range of possible values for the cosmological constant, including one that matches Weinberg's prediction and could create the observed accelerated expansion.[12] "String theory would provide a microscopic realization of what Weinberg needed," says Polchinski. There may be as many as 10^{500} possible configurations—10^{500} equally valid solutions to string theory's equations, and thus 10^{500} possible forms our universe could have taken. These possible universes would not just have different cosmological constants but would have different particles and forces, and maybe even different physical laws too.

Physicist Leonard Susskind, at Stanford University, coined the phrase "string landscape" to describe the gargantuan range of possible places, or energies, a universe might find itself in, depending on the parameters of its hidden dimensions. But why, out of almost infinitely many possible universes allowed by string theory, was ours coincidentally the only one lucky enough to be instantiated? At first glance, it seemed to make the puzzle of our serendipitous existence even more acute.

But Bousso saw a resonance with eternal inflation, laid out in Chapter 5, which churns out an ever-increasing number of bubble universes. Put the two together and suddenly you had a physical multiverse with different universes—and potentially each one could wrap up its extra dimensions in different Calabi-Yau manifolds, populating them with entirely different physical laws and properties. Where Guth once said that the universe is the ultimate free lunch, Linde now adds, "The universe is not just a free lunch: it is an eternal feast where all types of dishes are served."

The string multiverse gives new legs to the anthropic argument. All universes that can be born are born. Our universe may be a highly unlikely place to end up in, but if all possible universes are physically realized, then there's little point in worrying about how we ended up here, in this one, as we couldn't very well have found ourselves in one that couldn't support life.

The fine-tuning argument for the existence of God also seems to dissolve in the face of the string multiverse. We no longer need a divine being to make us a universe to order; string theory and eternal inflation can give us whatever we need instead. Writing in the *New York Times* in 2005, Christoph Schönborn, the archbishop of Vienna, denounced the multiverse hypothesis, alongside neo-Darwinism (the assertion not only that evolutionary processes created the variety of life on the planet but that those processes are random, purposeless, and unguided), stating that these scientific claims were "invented to avoid the overwhelming evidence for purpose and design found in modern science," whereas "the Catholic Church will again defend human reason by proclaiming that the immanent design evident in nature is real."[13]

Polchinski believes that Schönborn was right to feel threatened. "The multiverse is very much like natural selection: it's explaining how something that appears to be designed can come about naturally," he says. Polchinski admits, however, that it took time for him to be won over to the anthropic merits of the string landscape. While he and Bousso agreed totally on the scientific result they had found, Polchinski initially hit a psychological roadblock and told Bousso to soft-pedal the implication of their findings in their paper, which was that there was no point searching for a deep explanation for why the constants of nature take the values they do, as they simply are that way, the result of a random occurrence. Partly he felt a pang of fear, because physicists have been entrusted with the resources to build multibillion-dollar particle accelerators to measure the fundamental properties of physics, but if anthropic reasoning is correct, there's nothing much more to say about the measurements they find, such as the mass of the Higgs boson, other than they just happen to be that way, without any profound, deeper explanation. "It's a tremendously discouraging thing, and I didn't want to be the person to be so discouraging," Polchinski says.

Polchinski happily notes that his initial fears were unfounded. The string landscape has now had more than fifteen years to percolate into people's consciousnesses, and it has not led to the shutdown of the LHC. That is partly because what string theory took away with one hand, it has given with the other, providing a host of new and exotic things for physicists to search for at the LHC instead. The most interesting for our purposes is the search for mini black holes, which would stand as evidence for string theory's extra dimensions. In Chapter 9 we'll find out more about this hunt and whether particle physicists may be able to distinguish between a mini black hole and a baby universe produced in the lab.

But before that, we have to overcome the final hurdle to making a baby universe in the lab, in theory, at least. Polchinski and colleagues had shown that it is possible in principle to create a universe without the need for an initial white hole singularity, as Guth and his team had found independently. But that finding helped only if you were prepared to wait, possibly for eternity, for a quantum process to bring the baby bubble into existence. Universe makers needed a way to speed things up, and for that they needed to find a cosmic seed. The quest for this will take me to Tokyo, to hunt for the monopole that set Guth on the path to discover inflation in the first place.

8

The Little Bang Theory,
or How to Grow a Universe from a Monopole

Nobuyuki Sakai stands on the stage, legs slightly apart, in the traditional opening stance dictated by kendo, the Japanese sword-wielding martial art. He doesn't actually have a sword in his hand, though; what he does have is a laser pointer, to allow him to highlight aspects of the PowerPoint presentation displaying equations of gravity on a screen behind him. He shifts his weight back and forth between his feet, demonstrating the change in balance afforded by flexing the back foot at the ankle rather than lifting the heel off the floor, and then jabs forward, suddenly in attack mode.

It's slightly surreal watching Sakai, a bespectacled astrophysicist in his late forties who is decked out in fairly typical "physics clothes"—a sensible pair of beige trousers and a smart maroon shirt covered in pictures of clocks—in full kendo flow at this meeting at the Nippon Sport Science University in Tokyo, Japan. Stranger still is that, if anything, Sakai is underdressed in this garb; everyone else at this lecture (all males, I notice) is wearing a solemn black suit. Sport science is certainly far more formal than the average physics conference.

Sakai is presenting his research on optimizing kendo coaching techniques, using a strategy he has developed that relies on physics equations. Of course, I haven't traveled all the way to Tokyo to talk about martial arts. His day job involves applying those equations in more conventional realms in astrophysics and cosmology. "People often ask me why I study sports as well as the universe," he tells me after his presentation, laughing. "My answer is I do research in different areas just because these are the things that catch my interest. I don't stick to one thing."

I'm here to find out more about one of those other pursuits, which first enraptured Sakai when he was a graduate student in the 1990s and never let go, turning him into an avid universe maker. In 2006, Sakai and his colleagues laid out the most thorough blueprints yet for building a cosmos in a particle accelerator—his little bang theory, if you will—using ingredients that can, at least in principle, be made on Earth.

Until now, we've talked about how conventional physics says that if you could get hold of a little patch of false vacuum, you could set it inflating in the lab, transforming it into a whole new cosmos of its own. But I have skated over the question of how physicists might get hold of this crucial patch of false vacuum in the first place—and it's time to tackle that now. Can you make this universal germ? Or seek one out? Sakai and company were the ones who came up with an answer. They identified a candidate particle that could be used to seed a new universe, a magnetic monopole, which, they argue, is primed to inflate under the right circumstances. Just grab yourself a monopole, stick it in a particle accelerator, hurl things at it at high speed, and boom—you have a homemade universe. Simple.

The catch? No one is sure that magnetic monopoles exist or, if they do, where to find one. So I am here in Tokyo to ask Sakai more about how he came up with this idea and why we should have faith in ever finding monopoles at all. Over the decades they've been sought in moon rocks, in cosmic rays from space, under the ocean, and deep within Earth's crust, always in vain. But more recently things took a brighter turn after a set of lab experiments generated quasi-monopoles—entities that have properties suspiciously close to real monopoles—in a special kind of crystal. Meanwhile, the hunt for monopoles made at the Large Hadron Collider located near Geneva, Switzerland, has kicked up a gear. Does this give us reason to hope that we are nearing the announcement of the birth of our first baby universe, from a newly made monopole?

"As a mathematical subject, the creation of a universe is easy," says Sakai as we walk to a restaurant during a lunch break in the martial arts meeting. But in that regard, it's deceptive, kind of like taking up kendo as an adult, as Sakai did a few years back, he explains. "First it looks easy, but it isn't," he acknowledges. Putting something into practice—be it wielding a sword or birthing a baby cosmos—is, of course, a lot more challenging.

I say that we are "walking" to the restaurant, but it would be more accurate to say that we are wading, as torrential rain is beating down and the streets are flooded with about six inches of water. A freak typhoon has hit Japan, leading to this unseasonable weather. As we take shelter in the restaurant, Sakai, who is consummately polite and considerate, apologizes to me for probably the fourth time that day for the storm. I try to keep reassuring him that I really do not hold him personally responsible for the forces of nature, but I appreciate his courtesy.

As we settle in, Sakai tells me more about his childhood. We are hampered by my shameful ignorance of Japanese, and although Sakai is pretty fluent in English, we have to resort to Google Translate on Sakai's laptop to aid us with more obscure terms. Sakai was born in quiet Niigata prefecture, where his father was a train driver. While some generations back his family were practicing Buddhists, Sakai was raised without any reference to faith or spirituality. "Nothing, nothing at all," he says, shaking his head decisively and slicing the air with his hand for emphasis when I ask if he grew up in a religion. "That's common in Japan, I think."

As a child, Sakai had little training in martial arts and did not enjoy them; his fascination with kendo would arise only later. But he did love mathematics, which he began to study as an undergraduate at the Tokyo University of Science, before changing his mind and reapplying to enter a natural science program. "I like mathematics, but mathematics itself is abstract, and I want to understand the real world," Sakai explains.

Sakai first learned of the universe building project in the 1990s, while he was working toward his master's degree with astrophysicist Kei-ichi Maeda, now at Waseda University, in Tokyo. As Sakai talks me through this history, I suddenly realize how focused the narrative I have presented about the development of inflation in this book has been on the United States and Russia, as indeed was my own cosmology education at Cambridge and Brown—with particular emphasis on the innovations of Alan Guth and Andrei Linde. Sakai, however, was raised with a different yet equally valid view in which inflation is very much a tale of Japanese ingenuity.

In Chapter 3 I presented Guth as the father of inflation—which is certainly the mainstream view, and well deserved—while in Chapter 4 I discussed the crucial work of Linde in turning inflation into a workable theory. I have also noted that many cosmologists and physicists have played a role in its development, and I have not attempted to do them justice by crediting them all here. Rather, I have been concentrating on the researchers who not only helped to develop inflation but went on to engage in universe building in some capacity.

One of the names that I glossed over is that of Katsuhiko Sato, now the president of the National Institute of Natural Science and emeritus professor at the University of Tokyo, who in 1981 proposed a scenario in which the expansion of the infant universe rapidly accelerated.[1] Ring any bells? The process would be catchily dubbed "inflation" by Guth, who independently developed the same formula. But, as Sakai points out to me, Sato submitted his paper for publication before Guth, yet inflation has now become synonymous with Guth, while Sato's contribution is less well remembered globally. I have to admit that I am quite embarrassed about my ignorance of Sato's findings. "Japanese people sometimes complain that some good works by Japanese researchers were ignored in the world," says Sakai.

In 1982, Sato and his colleagues published a paper predicting that inflation could naturally lead to a multiverse of parallel cosmoses, anticipating the later developments of the inflationary multiverse, laid out in Chapter 5, by Alex Vilenkin, Linde, and many others.[2] As the Japanese team poetically put it in the abstract of their paper: "Although the Creator might have made a unitary universe, many mini-universes are produced sequentially afterward." In this sense, Sakai notes, "the idea of a child or baby universe was proposed very early by the Japanese group."

One of the authors of that paper was Sakai's master's supervisor Maeda, an expert on the burgeoning theory of inflation. He assigned some of his papers, along with others on the foundations of cosmology, to Sakai to read. Baby universe production wasn't specifically on that menu—perhaps it was deemed a bit frivolous to be a starter project for a serious-minded graduate student. But while Sakai was conscientiously combing through the references on his official reading list, he came across Guth's 1987 paper, with Eduardo Guendelman and Steven Blau, that claimed that, under very particular conditions, it may be possible to set off inflation in the lab, creating a baby universe, as we learned in Chapter 6.

From that moment, Sakai was hooked. Over the past two decades, while producing a steady stream of papers across a range of more conventional physics topics—and a sideline in the physics of kendo—he has attempted to develop a series of techniques to bring that wildly speculative idea to practical fruition. The search had begun for the key missing ingredient, the cosmic seed.

A clue to the form of that seed was sewn into both Guth's and Sato's proposals of inflation back in the 1980s. Both physicists had presented their models of a rapidly expanding universe as the solution to a handful of major cosmic conundrums that had been flummoxing physicists at the time. Recapping from Chapter 3, the most interesting of those puzzles, for our current purposes in this chapter, was the mystery of the missing magnetic monopoles.[3] In everyday life, we are familiar with magnets that come with two poles, north and south; Earth has its own magnetic dipole, as do all fridge magnets, and any other magnet you happen to chance upon. If you try to split one of these in half, you'll just end up with two smaller dipole magnets, each with a north and south pole, rather than with two separate poles.

Monopoles, however, are hypothetical particles that were first proposed to exist by British physicist and Nobel laureate Paul Dirac, who is one of the founders of quantum theory. In the 1930s, he proposed that it is possible to have a single magnetic charge—either a north pole or a south pole in isolation—just as we commonly find single particles that contain a single positive or negative electric charge, such as protons and electrons. Dirac had been developing a quantum description to explain why electrons always have the same negative charge wherever they are—never 10 percent less, half as much, or any other fraction of that charge. Since electric and magnetic effects are tied together (as we saw in Chapter 1), he felt that a theory of electric charges was crying out to include magnetic charges, or monopoles. Indeed, he calculated that if monopoles with a single stable magnetic charge do exist, they could stabilize the single electric charges too, explaining why you never find particles with a fractional electric charge.

If that wasn't enough, physicists have also invoked monopoles in theories to explain other aspects of particle physics, such as why the elementary particles called quarks cluster together within atomic nuclei rather than roaming free, and how protons can decay. So monopoles are certainly very handy little objects, at least in theory. But while there's very good reason to think that Dirac was right to posit them, so far none have ever been detected, puzzling physicists for the better part of a century.

The absence of these monopoles became even more pressing in the 1970s. Physicists at the time were excited about the possibility of explaining how the electromagnetic force and the strong and weak nuclear forces all arose from one joint force in the early universe, as described by a grand unified theory. Sato, like Guth, had investigated various attempts at constructing such frameworks in some detail, and both of them knew that if these models were right, then monopoles were an unavoidable byproduct. Not only that but, recapping from Chapter 3, Guth's calculations showed that if a grand unified theory held true, monopoles should have been produced in such abundance soon after the big bang that we'd be tripping over them now. And yet none were to be found anywhere. Why not?

Inflation had an answer. If the universe had expanded at a rapid rate in its youth, then it was possible that a small patch of the universe containing just one monopole could be stretched to span the whole of our observable universe today. So while our infant universe may well have been riddled with monopoles, it was perfectly possible that they had been so diluted by inflation that the chances of us spotting one now were slim.

But while inflation appeared to explain away the missing monopoles, in an ironic twist cosmologists in the 1990s began to think that these elusive particles might actually play a vital role in driving inflation. Remember that despite the fact that there is plenty of strong observational evidence that supports the predictions made by inflation—particularly, as we learned in Chapter 3, the ripple patterns spotted in the cosmic microwave background radiation left over from the big bang—and it is pretty much the mainstream cosmological textbook theory, the paradigm still lacks definitive proof (depending, of course, on whom you ask). There is still a big question mark hanging over inflation because there's no consensus as to why our universe began to inflate in the first place, though a number of possible mechanisms have been put forward.

It was while reading up on this issue of how to kick-start inflation that Sakai was reintroduced to monopoles, through a paper written in 1994 by Linde.[4] Never one to rest on his laurels, despite having invented new inflation and eternal chaotic inflation (as we learned in Chapters 4 and 5), Linde had come up with yet another flavor of inflation. I won't go into the details of this new mechanism, dubbed topological inflation, which Guendelman and Vilenkin also worked on, because it's not currently very popular.[5] But the key insight that Linde had—which Sakai picked up on—was that under the right conditions monopoles have inflationary superpowers. As Linde once put it to me: "The monopole is like God, producing space all the time."

That's a pretty mind-blowing statement. To see why monopoles have this ability and other objects—electrons, protons, acorns, ping-pong balls, or anything else you can think of—don't, we have to remember that monopoles are peculiar, not so much things as spherical knots in space. As we saw in Chapter 3, monopoles form as the temperature of the early universe drops. The cosmos is pervaded by a Higgs field, the field that gives particles their mass. As it cools, this field changes phase, just like steam condensing to water and then freezing to ice. At each juncture, the Higgs field endows otherwise indistinguishable particles with new properties, promoting some to the ranks of massive particles, such as electrons and quarks, while leaving others, such as the photon, among the massless also-rans. At the same time, according to grand unified theories, it's thought to peel apart a single master force to give us the separate electromagnetic, weak, and strong forces that we experience today.

Also recall from Chapter 3 that these phase changes have a side effect: they leave behind fractures in space, just like the cloudy defects that appear when water freezes, marking the boundaries between ice crystals. A monopole is the equivalent of a bubble in the ice, marking a misalignment in the direction of the Higgs field as the field sets around it. You can picture the monopole as the metal ball on the end of a medieval mace, with spikes sticking out of it, but here the spikes represent magnetic field lines pointing radially outward.

Linde realized that because the monopole isn't a particle in the conventional sense but a by-product of the way that the cosmos is configured around a point, it won't react to inflation in the typical way. As the universe rapidly expands, most particles get carried along for the ride. Not so the monopole, however, which inflates too. From the outside, to anybody holding the monopole in one hand, say, it would look like a point—a magnetically charged mini black hole, in fact. But from the inside it would generate its own space, growing ever larger.[6]

Linde took the notion a step further to come up with the boldest explanation yet for why we haven't stumbled on a monopole. It's not that monopoles are rare and tucked away in some remote unexplored cosmic crevice. Rather, a monopole has been hiding in plain sight all around us. We are living, Linde conjectured, *inside* a giant monopole.[7] And if a pointlike monopole happens to be hanging around inside our universe with us, then it's also pregnant with inflationary possibility, primed to expand when triggered and gestate a new universe of its own. The outcome is a never-ending series of inflating monopole universes, each embedded within the next, just like a nested set of matryoshka dolls.

There was no physical evidence for these matryoshka universes, but Sakai loved the idea. As a postdoctoral researcher in 1996, he ran computer simulations to show that monopoles were indeed engines that could drive inflation, confirming that they would create a consistent spacetime with the same properties, dimensions, and flow of time as our own.[8] In 1999 he followed up with a paper showing—to the relief of any still worried about the possibility—that any such baby universe inside a monopole would definitely be divorced from our cosmos, growing its own space rather than encroaching into ours.[9]

"The creation of the universe in a laboratory idea almost came up then," says Sakai. One of the reasons it did not, however, was that he was distracted by other things—most notably his search for a permanent position. This took him first to the University of Cambridge, in England, for another postdoctoral stint, in 2000, followed by a full-time job at Yamagata University, before moving on in 2012 to Yamaguchi University, where he remains today. As a young professor, however, he couldn't indulge in whimsical thoughts of cosmic creation until he had developed a body of more conventional research, so he focused on studying the patterns in the cosmic microwave background and the nature of compact rotating stars instead.

It was while at Yamagata that Sakai's new fascination with kendo was accidentally sparked, he tells me as we return to the conference for the afternoon session. A twist of fate saw Sakai's office temporarily relocated to the university's sports science wing while the physics building was being renovated. He struck up a friendship with a kendo professor in a neighboring office and decided to take up the martial art himself. Together, they then teamed up to pair physics know-how with kendo training to devise a novel coaching strategy. "Not many people in Japan enjoy sports, and the main reason, I think, is that many people cannot master the basic skills because there is no methodology for beginners," Sakai says. "Coaching is very abstract or intuitive, and that is what I want to improve."

As we take our seats in the lecture hall to listen to an hourlong plenary talk, I ask Sakai why—given his varied interests—universe making is something he has returned to time and again over the years. "It is difficult to construct a universe really, difficult to test the theory," he admits. "But it is the most fundamental subject: creation of the universe."

I am quickly reminded again of how different this sports science meeting is from a typical physics conference. Sakai is seated on my left, and the moment the lights dim, I notice the man seated to my right, as well as three students in front of me, rest their heads and settle down to sleep. A scan of the audience reveals that many others are taking this opportunity for a siesta too, and soon Sakai follows suit. I'm quite jet-lagged, so I figure I may as well have a quick snooze as well, since everybody else seems to be. The speaker is supposed to be talking about the long and illustrious history of martial arts training, but since I don't speak Japanese, it's wasted on me anyway. After the lights go up and this communal naptime is over, I ask Sakai if he really thinks that his calculations will change the way coaches teach kendo, which has been practiced for centuries and carries the weight of tradition. "It's a good point, so we shouldn't rush things," he says.

The same question, in some sense, applies in the context of his universe-building efforts. Will people feel comfortable thinking about an idea as radical as making a new universe, or is it too much of a break with traditional values? In 2006 Sakai and his colleagues finally published their scheme, placing the final puzzle piece in the jigsaw of ideas mentioned in the previous chapters, raising the project to a new level of practicality—and possibly to an accompanying level of concern.[10] Their calculations show that if a monopole is struck with enough energy in a particle accelerator, it will inflate to create a new universe. This baby cosmos will be masked from us, hidden within a microscopic black hole, but—as Blau, Guendelman, and Guth conjectured thirty years ago—it could pinch off from us and survive to evolve its own fully fledged cosmos, replete with stars, galaxies, planets, and people.

It was this paper that first caught my eye when I was a journalist at *New Scientist* looking for a snappy new story to write. It certainly had an attention-grabbing title (far more so than its predecessors in the literature on universe making, which tended to veil their hubris in technical jargon): "Is It Possible to Create a Universe out of a Monopole in the Laboratory?" But Sakai notes with a laugh that the paper was originally submitted with a stronger, more provocative title: "The Universe out of a Monopole in the Laboratory." The journal apparently changed it, softening it to a speculative question rather than a definitive statement, without the authors' knowledge or approval.

"Do you think that they wanted to play down the claims because it's too controversial to state that a universe could be made in a laboratory?" I ask him. "Or they thought that people might get overexcited or upset by the idea?" Sakai just smiles, lowers his head, and shrugs.

"Should people be worried about the possibility of making a baby universe, which could evolve life within?" I press on.

"Mmm," Sakai murmurs, and then pauses for a long time before he adds, slowly and solemnly, "In a sense we should worry about that possibility, but at the moment, there are still ambiguities—we still haven't discovered a monopole—so it's too far away. So far, monopoles cannot be detected or made, but experimenters are still searching for them."

It's worth noting that since writing that 2006 paper, Sakai has teamed up with Guendelman, who visited Japan to meet him, and together they came up with alternative ingredients to seed a universe in case we never find a monopole. But, if anything, those are even more exotic entities. For instance, the pair has calculated that if some of the string theory models we first heard about in Chapter 7 are right, then the universe may contain bizarre objects with odd energy and pressure properties that could be used to seed a universe.[11] "If we suppose unusual matter, we can create a child universe, and in string theory, many strange types of matter can appear," says Sakai. I won't go into detail about these alternative recipes since Sakai thinks there's greater reason to believe that monopoles, rather than these odd stringy bits of matter, will be found one day. Still, it's important to mention that the universe-building project doesn't stand or fall on the discovery of monopoles.

Okay, so if monopoles, for now, seem to be our best bet for making a baby universe, how do we get our hands on a monopole? As Sakai says, the search has been on for many decades, and later in this chapter I will be talking to a particle physicist at the LHC who's working to hunt them down. But first I'm going to call up a physicist whose job initially appears to have nothing at all to do with exotic particles, cosmic expansion, or baby universes. Rather, he studies the properties of artificial crystals made in labs around the world (far more cheaply than the ephemera generated in billion-dollar particle accelerators, I might add), which engineers hope might one day revolutionize the IT industry. It was while pondering some unexpected properties of one of these materials that physicists realized they could be harboring human-made magnetic monopoles locked inside.

※

Like many young physics students, Claudio Castelnovo was taught about the mystery of the missing monopoles, in his case at the University of Milan, in Italy. He knew that if these intriguing entities were found, it would bolster the quest to join together three fundamental forces of nature in one grand unified theory. And, like many undergraduates before him, after learning that, he promptly forgot about them. Monopoles were fun, but he had no plans to seek them out himself. He certainly never dreamt that within the next fifteen years he would win a series of awards for discovering magnetic monopoles in a newly engineered material called a spin ice.[12]

Castelnovo's passion wasn't particle physics or the early universe but a branch of physics that deals with condensed matter and studies how collections of atoms behave when they are locked together in materials and you change temperature and apply electric and magnetic fields. It's an area that has direct practical applications for building better electronic devices and improving computation. But more important for Castelnovo is that it offers a direct handle on the nature of reality at a tangible level. As he explains to me over the phone, the same elementary particles clump together in different ways to form innumerable materials with a myriad of physical properties; investigating all the nuances of grand unified theories, or even the standard model of particle physics, doesn't really leave physicists any the wiser about these immensely varied characteristics of everyday substances.

"To understand the properties of a piece of rock, or water, let alone understand life, we need to go far beyond knowing exactly everything about the elementary particles that make them up," says Castelnovo. "The question asked in condensed matter physics versus particle physics is, simply, rather than asking, 'What are the elementary constituents?' to ask, 'How do we explain what we observe?'"

With that broad question in mind, when Castelnovo embarked on his first postdoctoral position at Oxford University, in 2006, he chose to investigate the quirks of a newly engineered crystal called a spin ice. Material scientists hope to one day tune the electric and magnetic properties of such artificial substances to create more efficient computer hard drives and other devices.

As noted in Chapter 2, spin is a quantum characteristic of atoms and other particles. It can be pictured as a little arrow stuck on each particle that points up or down, corresponding to how the particle moves in a magnetic field. When all the spins belonging to the atoms of a material are aligned, the material itself becomes magnetized. That explains the "spin" part of the spin-ice moniker. The "ice" part is a reference to the fact that at low temperatures the material has a regular crystal structure, with every group of four neighboring atoms making the corners of a tetrahedron—similar to the pattern in which H_2O molecules set when water freezes.

Spin ices have some eccentricities that captivated Castelnovo's curiosity. For instance, when you cool many materials down to almost absolute zero (the lowest temperature possible, -273.15 degrees Celsius), their spins align and they become magnetic. Spin ices somehow resisted that fate, however. Imaging revealed that even at low temperatures their spins remained haphazard.[13] Yet, mystifyingly, the material still managed to display some degree of magnetism. How was that possible?

Castelnovo and his colleagues began to wonder: if normal spin alignment couldn't explain the origin of the material's magnetic powers, was there something else in the material that could? Experiments carried out by others had revealed what appeared to be regions of unexpectedly intense effects—magnetic sources and sinks. His team realized that the simplest way of interpreting them was that the material contained separated single magnetic charges—oppositely charged monopoles. These monopoles were emerging thanks to the pyramidal configuration of the atoms. As the material was cooled, in some places the spins at three of the four corners of a pyramid would align, either up or down, creating a net north or south magnetic charge at the pyramid's center. The pyramid of atoms then acted like a monopole.[14]

These manufactured monopoles are called quasiparticles because they can exist only within the spin ice—and they arise only because of a peculiar configuration of the atoms and spins in that material. You can't reach in and pluck one of these monopoles out of the material; it simply doesn't exist without those surroundings. By contrast, you could peel off an electron from one of the atoms in a spin ice, because electrons are real particles, not quasiparticles. But within the confines of the material, Castelnovo and his collaborators discovered, these monopoles could move about freely. Further experiments have shown that if you stick the spin ice in a magnetic field, the north charges move to one side of the material and the south charges move to the other—like boys and girls separated on either sides of a hall at the start of a particularly awkward school dance. "Then you turn off the magnetic field, and they relax in a way that pointlike objects would do, walking towards one another," Castelnovo explains.

How these monopoles are able to "walk" around is a bit of an odd thing to wrap your head around when you remember that the atoms themselves are locked in place within the spin ice. The atoms cannot wander anywhere—the only freedom they have is that their spins can flip from up to down and vice versa. So how can the monopoles, which appear to be tied to the atoms for their very existence, exhibit this luxury of unrestricted movement?

There's an everyday example that helps illuminate the origin of the monopoles' "motion." Think of a sports stadium filled with spectators performing the wave. We know that this wave is generated by tens of thousands of people in turn standing up, lifting their arms, lowering them, and then sitting down. Each person is responding to the actions of their nearest neighbors, signaling to the person when to rise. Imagine viewing this scene from far enough away that you can no longer make out the individuals in the audience, but just see the wave moving—a weird monster made of multiple hands that races round and round the arena. In reality, there's no real hand monster; it just appears that there is one thanks to the behavior of masses of people, who always remain at their seats. In the case of monopoles, it's the spins on neighboring atoms that are signaled to flip, allowing the monopole to "walk" along. Castelnovo, in focusing on the "moving" monopoles, was tracking the monster and not worrying about the complex series of spin flips that created it.

Since then, a host of other experiments have been carried out that show monopoles zipping about in spin ices, verifying an idea that Dirac had about how they should line up next to each other to form a chain, with north poles and south poles alternating in what's called a Dirac string.[15] The power of such tests is that they confirm that magnetic monopoles are allowed by nature, and appear to behave exactly as physicists have envisaged since the 1930s. "Looking for particles in the universe is exciting," says Castelnovo. "But when we fail to observe them, finding anything that comes close to being such a particle in the collective behavior of a condensed matter system is equally exciting."

But—and it's a big but—because these are quasiparticles, not real monopoles, there's no hope of ever grabbing hold of one, placing it in a particle accelerator, firing stuff at it, and creating a baby universe. For that, we need to either capture a real monopole that's floating out in space or make one. To learn more, I called up particle physicist Albert De Roeck, whom I've known since I first visited CERN in 2009 and who is part of the MoEDAL experiment at the LHC, which is designed to search for monopoles created at the collider.

Experimentally, monopole hunters have two prongs of attack, De Roeck explains. "We can say, 'Let's see what monopoles the skies bring us, in cosmic rays'"—particles produced elsewhere in the universe that hit our atmosphere with high energy as they travel past us. "They could be very heavy monopoles created at the time of the big bang which pass the

Earth from time to time, and we can check for that," he says. These super-heavy monopoles are unlikely to be generated in a collider because they would require more energy than our current accelerators have the capacity to throw in.

People have been looking for evidence of monopoles that fell to Earth for decades, in rocks and even in shells retrieved from the ocean deep. "The thinking is at the dawn of time, during the creation of Earth, there may have been a shower of monopoles passing through," says De Roeck. As monopoles pass through materials—especially when traversing though magnetic substances—they lose a lot of energy and may even come to a halt. So there's a good chance they could get stuck in dense materials such as rocks. Some of the physicists now in the MoEDAL team have previously analyzed rocks from the Antarctic, where Earth's own magnetic pole could draw in monopoles. Underground searches, such as the MACRO experiment located within the Gran Sasso Mountains in Italy, have hunted for them specifically too. But so far no luck.

It's possible that flying monopoles are getting jammed up in our atmosphere and never even make it to the ground. This thought occurred to physicists, who figured that since the moon has no atmosphere, it could have taken direct hits from monopoles from space. Since moon rock has a high magnetic charge, monopoles could well have gotten lodged in its surface. But again, decades of studying moon rocks have yielded nothing.

That leaves De Roeck's second line of attack: attempting to produce monopoles during the collisions at the LHC. There are some versions of the electroweak theory—which brings together the electromagnetic force and the weak force that describes radioactivity, as discussed in Chapter 3—that say that monopoles could be made during smash-ups at the TeV scale (the level of energy at which the LHC is currently hurling particles together), which is enough to whet De Roeck's appetite. "That's the interest for the experiment we are doing," De Roeck says, "gambling on these electroweak monopoles."

De Roeck notes that the experimental search for monopoles is much cheaper than the billion-dollar hunt for the Higgs boson that we're used to hearing about from the LHC. It's low-cost because it's piggybacking on the collisions that are already taking place. It turns out that building a detector to check to see if a stray monopole is produced and comes flying out of the machine is pretty straightforward. All you need is a piece of plastic (in this case, 18 square meters of a specially developed substance, boringly called CR-39) hung in the cavern where collisions take place and left there undisturbed for a year at a time. Monopoles carry the equivalent of 68.5 times the single electric charge of an electron, dramatically outstripping the electric effects of any other particles that might be racing round the collider. This means that if one strikes the plastic, it will burn a hole in it, while no other particle would.

"So what we do is we put in the plastic and we go on holiday—or rather go to do other LHC work! And at the end of the year we collect the plastic," says De Roeck. At the time of writing, the plastic had just completed its first year-long run and was undergoing the long process of etching, a chemical treatment that will render any holes visible under a microscope. "If we find this hole, we know that a very nonstandard particle passed through it because no standard particle can burn a hole—the plastic just won't react," says De Roeck.

While waiting for these results, the team is also pondering placing 1,000 square meters of the same CR-39 plastic sheeting on a mountain in the Canary Islands to try to catch flying monopoles, if any are out there and make it to high altitudes unimpeded. "If some of these very heavy monopoles exist, we should see holes," says De Roeck.

The team also has another trick that they are employing in the LHC's tunnels. This one goes a step beyond just detecting the presence of monopoles that pass through the detector, striving to catch some and keep hold of them. This, of course, is exactly what a universe builder would need. Again, it's a simple test: the MoEDAL team has placed boxes with aluminum bars—literal monopole cages—in the cavern. If a monopole passes through, its magnetic charge should cause it to slow down and bind with an aluminum atom's nucleus, stopping it in its tracks and capturing it. The aluminum bars are then analyzed with supersensitive magnets to see if any monopoles have been caught in the trap.

Technically, the magnets used to search for trapped monopoles are called superconducting quantum interference devices, or SQUIDs for short—a nickname that inspires images far more interesting than the instruments deserve. When a north magnetic pole passes through a SQUID, it sets an electric current running in the device, but this current stops when it encounters a south pole. So if you pass an ordinary bar magnet, like one you might find stuck on your fridge, through a SQUID, it will only temporarily set off a current. But a north monopole in isolation would trigger a current that is not curtailed. "If there is a net current running at the end when we pass this material through it, then there must be a monopole in it," De Roeck explains.

Of course, if the team wanted to be really thorough, they could dismantle the whole of the LHC to see if a monopole has gotten stuck anywhere along its 27-kilometer ring. "But people don't want to do that—they say the experiment is too expensive, so they don't want to break it up," De Roeck laughs.

In July 2016, the monopole hunters presented their first aluminum cage results, from traps that were laid during the LHC's 2012 Higgs-discovery run (which ran at 8 TeV, just over half the energy of its current 13 TeV run), using 126 kilograms of aluminum. "Unfortunately, no monopoles were found," says De Roeck, "but that was the prototype of the detector." Traps containing about 800 kilograms of aluminum are now in place for the fully fledged experiment, which coincides with the LHC's higher-energy run that began in 2015. "We are of course anxious to see what's coming out of that," De Roeck says.

So if we want to follow Sakai's monopole-based recipe for cooking up a universe, we may still have to wait for all the ingredients to come together. This is not a bad thing, since De Roeck estimates that making a universe in a particle accelerator will require hurling particles at the monopole at energies beyond the capability of the LHC, and it may well take at least another fifty years before we develop the technology needed to reach those energies. This energy limitation, and the prospects for building future accelerators— or boosting the one we have—is something I will come back to in more depth in Chapter 9.

It's also an issue that I raised back in Tokyo with Sakai. He began toying with this subject some twenty years ago, as a student, before the LHC was up and running. Does the thought of potentially waiting for other accelerators to be built disappoint him? "Many researchers think science should be verified by experiments, and I agree," Sakai says. "But alongside that, we can imagine many interesting possibilities, and so even if a theory cannot be tested promptly, it's important to consider it theoretically."

I wonder if all this is nothing more than a thought experiment for Sakai, like Einstein musing on what it would be like to ride a light beam, leading him to come up with his theory of special relativity, which we discussed way back in Chapter 1. Is Sakai's passion for this project driven purely by the intellectual thrill of conjuring new ways to make a universe in theory? Or has he always felt that this was a project that really will come to fruition in practice?

"Do you think it really will be possible to make a universe in the laboratory one day?" I ask.

Sakai pauses, dips his head in thought, and then raises it again. His reply comes slowly but with determination: "Yes. Someday. Yes." He gives me a shy smile, knowing he may come across as a romantic dreamer. "I hope," he adds quietly. "I hope. But there are many hurdles." One of his biggest worries is not that we will fail to make a universe but that we might actually succeed but fail to notice. By his own calculations and those of others before him, from the outside a baby universe would masquerade as a mini black hole. LHC physicists have been poring through data for signs of these but so far have come up empty-handed. But even if they find one, how will we know if it's just a black hole or if it could harbor its own cosmos, its own civilizations? Sakai rests his chin on his hand and asks with a frown, "How can we even say that a universe is created?"

The answer could lie in the death throes of a black hole. It was long thought that these cosmic crunchers, once formed, were permanent fixtures in the universe, hoovering up unsuspecting objects that veered too close. But in the 1970s Stephen Hawking made the radical suggestion that they actually spew out particles. In the next chapter, we'll meet one of his students, a black hole expert, and investigate whether these emissions of particles, called Hawking radiation, could provide the last communiqué from our baby universe before it ends transmission and leaves home forever.

9

Postcards from the Edge: Detecting and Communicating with Our Baby Universe

When American physicist Don Page was a young researcher, more than thirty years ago, he found himself at the foot of Mont Blanc, the highest peak in the Swiss Alps. He was in the region for a physics conference and, during one of the breaks in the meeting, was afforded the chance to climb the mountain. Page knew that this was a dangerous expedition, and his mathematical mind began to calculate the odds of death. He figured that maybe ten thousand people braved the ascent each year. From reports, he'd heard that about ten of those died during the attempt.

Many people would tot up the numbers, calculate that the chance of dying was roughly one in a thousand, and then weigh up whether that was small enough to dismiss. But Page has a different way of looking at the world. He advocates the view of quantum theory that broadly says that whenever you make a decision, reality splits into parallel universes, each containing alternative versions of yourself, with your cloned selves enacting different choices in different worlds. That meant that Page could not simply ignore even a tiny risk; if he chose to climb the mountain, then in some universe he was inevitably condemning one version of his future self to death.

Page's deliberations went thus: At that age, he estimated that his natural life span would give him roughly another fifty years, or twenty thousand days on Earth. By choosing to tackle the mountain, reality would split to create a thousand clones of himself, each with around twenty thousand days left on the clock. One of these would die that day. Crudely speaking, this meant that, after averaging out that loss over the whole ensemble of "Don Pages" who would pop into being across the multiverse when he began his ascent, this would work out to be the equivalent of reducing the life expectancy of any one of those Don Pages by twenty days. Imagining the exhilaration as he reached the highest climbing station, Page reasoned that, on balance, it was "well worth losing twenty days," and so he made the decision to climb up.

"Maybe I didn't do this calculation as sophisticatedly as I should have," laughs Page, adding that back then he was a single man, which afforded him the luxury of being reckless. "After being married, with responsibility to my family and my wife, I wouldn't have done it," he adds. Causing loss and pain to even just one in a thousand of the alt–Don Page's families would be a price that was too heavy to pay.

I've called up Page—in this version of reality, at least—for a few reasons. One is that his specialty is black holes. Now based at the University of Alberta in Edmonton, Page cut his physics teeth back in the seventies, first as a student and then as a collaborator of Stephen Hawking's. Together they worked on Hawking's surprising discovery that black holes are not completely "black" after all. It had been thought that these cosmic objects swallowed up everything that came too close, including light itself, forever. Instead, Hawking realized, black holes can spit out radiation in the form of light and particles, which, in theory, can be detected by astronomers.

Hawking radiation may be the key to solving the major problem that I hope to address in this chapter: the budding universe builders we've met so far say that if we do eventually make a baby cosmos in a particle accelerator, then to the physicists peering in, it would look for all intents and purposes like a mini black hole. I'm slightly anxious that this means that even if we made a child cosmos, we wouldn't actually ever know we had succeeded—a concern that Nobuyuki Sakai had also raised (see Chapter 8)—instead mistaking it for one of the mini black holes that particle physicists have long been hunting. That would certainly be an anticlimax to the whole project.

But a few researchers have proposed that a baby universe can be distinguished from a mini black hole, even from the outside, by looking closely at the Hawking radiation emanating from it. Both are predicted to spray out a shower of particles, like tiny Catherine wheels. It's possible that subtle differences in this fireworks display could herald the birth of our own child cosmos.

I also have another motive for speaking with Page that goes beyond his expertise on black holes, though. I first encountered him in 2007, when I attended a physics conference at Cambridge University organized by Hawking to celebrate inflationary theory.[1] If you have a long memory and still recall Chapter 3, this was also the first time I saw Alan Guth, father of inflation, where he was happily dozing through a lecture (not the one that was given by Page, I hasten to add). Page, by then approaching his sixties, with a fluffy gray beard, glasses, and thick curls of dark brown hair, began speaking about aspects of the cosmology of the very early universe at his trademark breakneck pace.

At this point, I recall few of the details of his talk. What I do remember vividly, however, was eavesdropping on the conversation of two Cambridge students seated in front of me. As soon as Page started speaking they began to scoff. "You know Page believes in parallel universes?" one whispered to his pal. His friend responded equally derisively: "And he's an evangelical Christian!" They snickered together though the opening minutes of Page's talk before standing up, pushing past the others seated next to them, and leaving the auditorium. They had evidently decided that someone who openly held such bizarre beliefs didn't deserve their attention or courtesy.

I felt troubled by this display of contempt toward Page. Yet I also had to admit that his devout Christianity coupled with his sincere adherence to the existence of many alternative selves makes for a seemingly oddball mix. I first mentioned the parallel universe theory that arises from one interpretation of quantum theory in Chapter 2, where I also discussed why it might be tough for religious people to subscribe to this particular scientific view. While one version of you may be a pious churchgoer, there may be another that is a hell-raiser. Why should your "bad" clone accept judgment from God for misdeeds that you were inevitably predetermined to carry out in at least one universe?

So my second reason for calling Page—in addition to asking about how we might be able to detect a baby universe in the lab—is to learn about his philosophical worldview. Does the task of making a baby universe in the lab threaten his sensibilities as an evangelical Christian because it treads on God's toes? Or, as someone who also believes in parallel universes, does he assume that if baby universes can be made by humans, they inevitably will be, in some version of reality? And can he make peace with that?

Finally, the notion of Hawking radiation has profound implications for the life span of our baby universe—or at least for how much time we will have in the lab to observe it before it escapes our reach. What does the physics of black holes tell us about this possible fate of our child cosmos? And is there any way we, as its creators, could control our daughter world, or communicate with it?

Page was born in 1948 and grew up in Alaska, where his parents taught elementary school in various indigenous villages. There were few facilities for those who wanted to pursue education further; for high school, kids either had to move away to a boarding school or study by correspondence course. Page opted for the latter and admits this led to him growing up in a very "sheltered environment." His parents were Baptists and taught Page about God and Jesus, but stressed to him the importance of making his own decision about what to believe. At twelve he chose to profess his faith at a small church in southern Missouri where his uncle was the pastor.

His first time living away from home came in 1967, when Page, who had always had an aptitude for math, headed off to pursue the subject at William Jewell College, a Christian undergraduate institute in Liberty, Missouri, which his parents and grandfather had also attended. Once there, he found the college's physics program was much better than its math program, and that this was a science where he could use his knack for math, so he switched focus. Inspired by a campaign led by evangelical preacher Billy Graham that was passing through the town, Page also rededicated his life to evangelical Christianity at this time. "This wasn't really a change of direction for me, but an affirmation that I am now living on my own and I wanted to reaffirm my faith," says Page.

For graduate study, Page headed off to the California Institute of Technology, in Pasadena, to specialize in physics and astronomy. His PhD adviser was Kip Thorne, one of the masterminds behind the wormhole theory we encountered in Chapter 6 (recall that wormholes are the physical tunnels that would connect a baby universe to our own spacetime). It was at Caltech that Page first met Stephen Hawking, then a young researcher in his early thirties, who, as discussed in Chapter 1, had already made a name for himself through his work studying singularities—the infinitely dense points of matter thought to lie at the heart of a black hole and at the origin of our universe.

A couple of weird black hole findings that Hawking had already been developing caught Page's attention. So far in this book we've discussed two types of black holes lurking in our universe: the stellar-mass kind and the supermassive kind. As described in Chapter 6, the first are formed from certain stars with masses bigger than about five times the mass of our Sun, but less than roughly a hundred times our solar mass. When these stars reach the ends of their lives, having burned all their fuel, they collapse down. The conventional view is that all their matter is crunched into a infinitesimally small singularity. Supermassive black holes can be between hundreds of thousands and billions of times more massive than our Sun, and how they form remains a mystery, one that we'll return to in this chapter. But in the early 1970s Hawking speculated about another, far tinier cousin to these cosmic beasts: the primordial black hole.

Hawking noted that in the first microsecond after the big bang, the cosmos would have been dense enough that mini black holes, just 10^{-35} meters across (the so-called Planck scale, which we encountered in Chapter 7) and weighing in at about 100 million kilograms, might be produced. As more time passed, heavier primordial holes could have joined them, and these could have been left peppered around the Milky Way today in huge numbers.

It's still not clear whether or not these mini monsters exist out in space. And at first it seemed pretty unlikely that astronomers would ever be able to spot them, even in theory, if they were out there. These were minuscule objects that—by the definition of what it meant to be a black hole—did not give off any light for telescopes to pick up. After all, conventional wisdom said that nothing, not even light, can escape a black hole's clutches. But before his arrival at Caltech, Hawking had already begun challenging that mainstream view. Shockingly, his early calculations suggested that black holes actually leak particles. It's this hypothetical Hawking radiation that could enable telescopes to track down primordial black holes in the sky and—more important for our purposes—could aid the search for both mini black holes and baby universes born in a particle accelerator.

To understand how Hawking arrived at this startling result, we need to quickly recap the state of black hole affairs when he first turned his mind to the question. As we saw in Chapter 6, black holes pack so much mass into their cores that they dramatically warp spacetime around themselves. According to Einstein's equations, once an object passes too close, veering through the black hole's event horizon—an imaginary spherical surface encompassing the hole—it would be captured and drawn inexorably in toward the singularity at the hole's core. Once you cross the event horizon, in this view, not even light, the fastest thing in the universe, could escape.

Those calculations had all been done using the rules laid out by Einstein's theory of gravity, general relativity. But by the 1970s physicists had long been aware that to truly understand a black hole's heart you also need to think about quantum theory. That's because a singularity is a tiny point and, in small spaces, quantum laws reign. This was even more apparent in the case of primordial black holes, which were just a fraction of the size of even the smallest elementary particles and thus lived well within the quantum realm. But, as we saw in Chapter 7, physicists were struggling to meld Einstein's general theory of relativity with quantum mechanics.

By the early seventies, a number of researchers scattered around the world were puzzling over what might happen at a black hole's event horizon if you took quantum theory into account. It took Hawking to make sense of it, though, explaining that when quantum effects were factored in, black holes seemed to be far less formidable prisons. He has since come up with a handy way of viewing this process, which he outlined in his 1988 book *A Brief History of Time*.[2]

Quantum theory, as we saw back in Chapter 2, tells you that pairs of particles and their antiparticles—an electron and a positron, say—can pop into existence, morphing out of the vacuum at any instant. These are called virtual particles because they usually only get a short taste of life, dancing around each other, before they are drawn together again. When close enough, their antithetical properties cancel each other out, and they are annihilated, vanishing from reality—unless, that is, as Hawking realized, they are created near the edge of a black hole.

What would happen if, Hawking conjectured, this virtual pair of particles is born just outside the event horizon? In that case, there's a danger that one of the particles might veer across the boundary and be swallowed by the black hole, leaving its partner bereft outside. In the process, this lone particle left outside would suddenly be gifted with full existence—since it could no longer recombine with its partner—and would shoot away from the black hole. Quantum calculations said that this could occur repeatedly, creating a shower of newly widowed particles and photons of light whose spouses had been guzzled up by the black hole. This is what has become known as Hawking radiation.

At first even Hawking couldn't believe what his equations were telling him. "He thought it didn't make sense because he seems to be getting an infinite number of particles out," says Page. But that was because Hawking had assumed that the black holes remained there forever. In fact, as these particles radiate away from its surface, Hawking realized, the black hole gradually shrinks. The reason for this is that when one member of the virtual pair escapes from the edge of the black hole, imbued with a new real existence, it speeds off with positive energy. To compensate for this, as its doomed virtual partner drops into the black hole, it diminishes the energy of the black hole. To someone watching from a distance, the black hole appears to slowly get smaller as it emits particles, until one day it evaporates entirely.

Hawking also realized that you can think of a black hole spewing out particles as a bit like a hot coal giving off heat, and assign it a "temperature." Heavier black holes have lower temperatures, spewing out radiation extremely slowly, just as cooler objects in general lose heat less rapidly than hotter objects. A stellar black hole, for instance, might have a temperature of about a millionth of a degree Celsius—making it a tough ask for astronomers to pick up its feeble radiation signature. ("It's like trying to tune in to your favorite radio station if its signal was being broadcast from a planet light-years away," said Greg Landsberg, a hunter of mini black holes whom we will meet later in the chapter.) Such a black hole would also take something like 10^{64} years to completely shrink away, far longer than the age of the universe today.

Primordial black holes should be much hotter, however, and would evaporate far more quickly, as a result. The smallest, created soon after the birth of the universe, would probably already have spewed out their entire contents and should be long gone. But some could be in their last throes right now, and it is for signs of these that astronomers are currently hunting. As the hole shrinks, it gets hotter and hotter, spilling out high-energy particles and light faster and faster in an avalanche effect. In its last gasp, a primordial black hole could explode in a huge burst of radiation, with energies equivalent to a hydrogen bomb. (Don't worry—a mini black hole created in the lab would not release a dangerous amount of energy when it evaporates.)

Astronomers were keen to find these hypothesized primordial black holes not just because they were interesting objects in their own right but also because they could potentially solve another cosmic conundrum. It was possible that these tiny invisible concentrations of mass were the unidentified source of dark matter, the mysterious substance that seems to make up the vast majority of matter in our universe, as described in Chapter 7. One of the first papers that Page and Hawking wrote together, published in 1976, dealt with this idea.[3] It involved calculating how many primordial black holes might be out there to see if there could be enough to be responsible for dark matter's observed properties.

Unfortunately, Page and Hawking calculated that the maximum number of primordial black holes that could be out there and yet still remain hidden from telescopes was too low to account for all the dark matter that physicists believe must be lurking in the universe. But astronomers are still searching for them, in case they could make up even a fraction of the elusive substance. NASA's Fermi Gamma-Ray Space Telescope, launched in 2008, for instance, is hunting for Hawking radiation given out by primordial black holes in the form of gamma rays, a high-energy type of light, but has yet to find any evidence that these mini holes exist. Others have proposed that primordial black holes could be streaming through Earth all the time. Since they are so small, they would meet no obstacle, but they may leave an acoustic signal that could be picked up with ultrasensitive seismic equipment, the sort that detects early signs of earthquakes. (I also recall a rather amusing paper that suggested that such streaming primordial black holes should be blamed for crashing your computer as they waft through it. If that's the case, there seem to be many of them around.)

Page and Hawking's primordial black hole paper was written during Hawking's yearlong visit to Caltech in 1974–75. By then, Hawking was already afflicted with amyotrophic lateral sclerosis and was wheelchair-bound, but he had yet to receive the tracheotomy that would force him to use an artificial voice, creating the speech patterns he is so readily identified with today. Hawking became Page's informal PhD advisor, alongside Thorne. When Page graduated and was looking for his first job, Hawking offered him a place at Cambridge University, where he was based. There they would continue to ponder this bizarre radiation together.

It was at Cambridge that Page first encountered real hostility toward his religious beliefs from the academic community. He was a member of Darwin College—one of the university's thirty-one colleges—and recalls sitting on a committee where members were discussing whether to allow a minister at a local church the right to come and dine there on a regular basis. The members were reluctant, and Page recalls one of them not only saying that the minister should not be invited but also adding, "I won't be satisfied until the last Christians are eradicated from this college." Recounting the tale some forty years on, Page laughs raucously, but then adds that at the time, those words caused him to "shake a little bit."

Page found a far more welcoming reception from Hawking and his family. Rather than living in college, Page was also paid to help care for his mentor, a role that saw him living with Hawking, his wife, Jane, and their children for his three-year tenure as a postdoctoral researcher. Hawking did not share Page's faith, but he was open to listening to the younger man's thoughts on the subject—often puncturing any attempt by Page to preach with a witty repartee. "One time I talked about how Jesus freed a man from evil spirits and sent those spirits into some pigs, and then the pigs went into the sea," says Page.[4] "I told this story to Stephen and he said, 'Well, the Royal Society for the Prevention of Cruelty to Animals wouldn't like that story, would they?'"

On another occasion, Page recalls sitting with Hawking over breakfast and telling him how Jesus had talked about how people will be separated into those who go to heaven and those who don't. "I told him Jesus said that two people will be in bed, and one will be taken and one left, and there's two in the field and one will be taken, and one left, and then Stephen pipes up, 'And two at the breakfast table!'" Page laughs.[5] For Hawking, Page speculates, God—if there was one at all—could at best be identified with the laws of physics. But there was, in Hawking's view, no logical reason to posit the existence of a personal God who interfered with or cared about your actions.

Page relished the extra time with Hawking that his living arrangements offered him. On many occasions they had their most fertile professional discussions while pondering open puzzles during the walk to the department in the morning. "Once or twice I managed to come up with the answer before he did, but it was pretty rare," Page recalls. Though they had little reckoning of it at the time, one such paradox about black hole radiation that was up for discussion would lead physicists to contemplate whether black holes in the sky might house their own universes, and if so, how you could detect them. Some forty years later, the same ideas would inform the discussion of how to recognize whether a baby universe had been successfully made in the lab.

The paradox first arose when Hawking thought about what happened to the information about objects gulped down by a black hole. Here was the issue: Quantum laws say that information is never completely lost from the universe—it's always preserved in some state, though it may be garbled. For example, even if you burned a copy of this book, in theory a smart enough forensic scientist of the future, with enough know-how, would be able to glean enough information in the ashes to figuratively wind the clock back and piece it together again.

Until Hawking had come up with his notion that black holes spew out radiation, people had just assumed that if you threw a copy of a book into a black hole—though the book would be violently distorted by being first stretched by the black hole's gravity and then ripped apart—somewhere in the core of the black hole, jumbled in with the squished-up remnants of other objects, would be enough information to reconstruct the book again. But there was a problem when you factored in Hawking radiation.

According to Hawking's idea, as black holes emit particles, they get smaller and smaller until they eventually fade away. So what happens to the information about their contents when the holes vanish from the universe? Hawking's first thought was that the escaping particles might somehow encode that information, carrying it away with them. It's a notion he has flip-flopped on over the years. Back in the seventies, he calculated that the form of the radiation made this impossible because it was "thermal"— essentially it was like the heat given off when burning a book, which wouldn't be much help in telling you what the original book had been. Hawking was so confident that the information must be lost—though this contradicted quantum laws—that he and Thorne even made a bet to that effect with another quantum physicist, John Preskill, now at Caltech, who ardently believed that when physicists better understood what was going on, quantum laws would be found to win out. The prize would be an encyclopedia of the winner's choosing.

In 2004 Hawking, convinced by ideas from string theory that seemed to suggest that information might be carried off in the Hawking radiation after all, conceded that Preskill was right, and made good on their wager. But things have yet to be fully resolved. In recent years, physicists analyzing the situation in more depth have called that claim into question, and the jury is still out.[6] The black hole information paradox, as it has infamously become known, has yet to be definitively solved.

For our purposes, the most interesting thing about this paradox is one seemingly wild possibility that was first put forward back in the 1970s and has been bandied about by many physicists since. This is the idea that black holes serve as gateways to other universes, and so information falling through an event horizon is simply siphoned off into a baby cosmos. As long as quantum theory holds strong in that child universe, then quantum laws would not be violated, even though that information is inaccessible to us. The notion, however outlandish, is a cosmic counterpart of the plans for making a baby universe in the lab that we've been discussing throughout this book. Instead of trying to make a universe artificially, these physicists suggested that such child universes might occur naturally, inside conventional black holes. But is there any evidence to suggest that such large-scale alternative universes are hiding within black holes?

The answer is not yet. But there may be soon—and that would be good news for the universe-making project too. Over the years, the notion that black holes in the sky might harbor parallel worlds has been bolstered by a number of different theoretical models. In the 1990s, for instance, physicist Lee Smolin at the Perimeter Institute in Waterloo, Ontario, put forward an early version of a multiverse theory, in which each time a star collapsed to create what looked from the outside to be a black hole, it birthed a new universe, which could have physical parameters that were slightly different from our own. His motivation for proposing this was to explain why our universe seems to fortuitously have just the right parameters needed for human life to arise. In Chapter 7 we saw that physicists have struggled to explain this based on the laws of physics. Smolin dubbed his model "cosmological natural selection" because it provided a Darwinian-inspired answer: universes that are fit for producing stars are also more hospitable to life because carbon and other heavy elements necessary for our existence were forged within the furnaces inside stars. Such universes will, in turn, spawn more black holes, with similar life-friendly characteristics. This means that, over time, there will be vastly more universes conducive to evolving human life than not. Therefore, it should be no surprise that we find ourselves living in one of them.

More recently, those working on loop quantum gravity, a theory that I introduced back in Chapter 2, have come to a similar conclusion. Proponents of loop quantum gravity argue that spacetime is knitted together from tiny threads of energy and that these threads prevent spacetime from ever crunching down to an infinitely small singularity. Instead, they say, our universe has existed eternally. At some point in its past, the cosmos contracted, squeezing down to a tiny but finite size before bouncing outward again; we are currently living in its expanding, post-bounce phase. In 2013, physicists Rodolfo Gambini, at the University of the Republic, Montevideo, in Uruguay, and Jorge Pullin, at Louisiana State University, in Baton Rouge, argued that loop quantum gravity models also do away with the singularities thought to lie at the heart of black holes, again replacing them with wormholes—tunnels in spacetime that take those who fall past the event horizon to an alternative universe.

But for our purposes, the most significant such suggestion comes from inflationary maestro Alex Vilenkin, whom we first met in Chapter 5, who earlier came up with a way of conjuring a universe from nothing (or nothing tangible, at least). In 2015 he and his colleagues proposed that, based on inflationary theory—the mainstream framework that lies at the center of universe building—some supermassive black holes may well be gateways to parallel cosmoses too.[7]

The paper caused some serious excitement among cosmologists, not so much because of the universe-hiding prediction, which has been around in science and science fiction for decades, but because, as Vilenkin excitedly told me when he sent me a copy of the paper upon its release, it offers a "new way to test multiverse models observationally." Of course, I also realized that the paper is another boost to the idea that a mini black hole made in a particle accelerator might harbor a child cosmos, but with the added treat that we may be able to find evidence that these shrouded alternative universes exist—at least in space.

Vilenkin's paper suggests a mechanism for the formation of supermassive black holes in our universe that those who have stuck with the book thus far will find familiar. Vilenkin, with Jaume Garriga, of the University of Barcelona, and Jun Zhang, of Tufts University, argue that such gigantic black holes could have been formed by little bubbles of vacuum embedded in our early universe. These would have expanded during our infant universe's inflationary phase. If these bubbles were small, then when inflation stopped in our cosmos, they would have collapsed down to a singularity, creating a conventional black hole. But if they were heavy enough, their bubble interior would have continued to inflate, creating a new universe. To us, on the outside, the new universe would look like a black hole and would be connected to our universe via a wormhole. (I told you this would sound familiar.)[8]

In their model, all black holes above a certain critical mass must house parallel universes, while all those below this mass can only be ordinary black holes.[9] The physicists' early, rough calculations predict a certain distribution of black holes of different masses in the sky. They estimate that heavier ordinary black holes are much more likely to form than lighter ones, that is, their probability increases with mass. But that's only true up to the critical mass; beyond this mass, hidden inflating universes are generated, but these should be rare, so the numbers should plummet again. What's appealing about these ideas is that astronomers can scour the skies for black holes and tally up how many they find with different masses. If this distribution fits with the predictions Vilenkin and others have made, it will stand as evidence for the model and thus for the claim that black holes can hold parallel worlds.[10]

Okay, so where does this leave us? The math coming from inflation theory says that supermassive black holes, hanging out in the night sky, may well be portals to new worlds. And, particularly interesting for our purposes, the calculations suggest that astronomers might be able to find proof that these hidden universe gateways are real. But would baby universes created in the lab have their own calling card?

It seems that there's a good chance they would, as some physicists have proposed that baby universes might give out a distinctive Hawking-radiation signature. To explain more, I'm going to take a brief detour from Page and call up an old lecturer of mine, Greg Landsberg, a Russian particle physicist at Brown University, in Providence, Rhode Island. In 2001, Landsberg became famous (well, among physicists, at least) when he calculated that the soon-to-be-turned-on LHC could end up becoming what he likes to call a "black hole factory."[11]

✳

Landsberg, like many experimental particle physicists, is a busy man, with an international commute. Whenever I try to call him, it's a guess as to which continent he will be on, as he travels back and forth between Rhode Island and CERN's Geneva lab, where he has been conducting a search for any oddities that cannot be explained by the standard model of particle physics and which thus might point to the elusive theory of everything that physicists have been seeking for decades.

With blond, slightly receding hair and a round boyish face, Landsberg, still only in his forties, is a man who will happily seize hold of wild ideas and run with them, as evidenced by the fact that just over fifteen years ago he became one of a handful of physicists who made the notion of black hole fabrication at the LHC fashionable. But despite his openness to bold claims, he does not suffer fools gladly. (Just ask one of my fellow grad students at Brown who faced Landsberg's ire and got shouted down in our experimental physics class after he was cheeky enough to try to give an end-of-term presentation with completely blank slides, having run out of time to prepare them.) Landsberg has written extensively about how to detect mini black holes at the LHC, so he should be the right person to tell me if Hawking radiation really is a sensible way to peer beneath the shroud of a black hole's event horizon to find out if there is a whole new world behind the veil.

Landsberg "followed the family trajectory" into physics as an undergraduate in Moscow; his grandfather specialized in optics, while his father was a particle physicist too. He then moved to New York for graduate study at Stony Brook University, joining an experiment at what was, back in the early 1990s, then the world's largest particle accelerator, the Tevatron, housed in a tunnel at Fermilab in Batavia, Illinois. In Chapter 7, we saw that physicists have good reason to think that the standard model of particle physics, though extremely successful, cannot be the final word on the workings of the subatomic world. This has led physicists to posit a cornucopia of alternative models, some of which, like string theory, attempt to bring together quantum theory and Einstein's general relativity. At Fermilab, Landsberg began his first serious search for clues that might show up in the debris of high-energy particle collisions—unexpected particles, say—that couldn't be explained by the standard model and might hint at the form that the ultimate theory of reality should take.

Baby universes born in a lab can certainly be classed as "nonstandard," but what should particle physicists sifting through post-collision wreckage look out for that would mark their presence? Way back in 1991 physicists Kristine Larsen, at Central Connecticut State University, in New Britain, and Ronald Mallett, at the University of Connecticut, in Storrs, wrote a paper inspired by Steve Blau, Eduardo Guendelman, and Alan Guth's 1987 publication, which was the first to propose manufacturing a child cosmos. Blau, Guendelman, and Guth had noted that from the outside, their child universe would masquerade as a black hole, and Larsen and Mallett decided to examine the Hawking radiation it would give out.

As we've seen, in theory, the smaller the hole, the higher its temperature, meaning that it sparks more light and particles and is thus easier to see in the dark. Larsen and Mallett used this size and temperature relationship to think about how the rate of evaporation of a baby universe might differ from that of a mini black hole. As a black hole shrinks, it should get hotter and hotter, emitting radiation at an ever faster rate. At the very end, just before vanishing entirely, this would create an unmistakable fireworks display of particles exploding outward. If such an event occurs within a particle accelerator, physicists would recognize it by looking at the speed, direction, and type of daughter particles pouring out of the black hole. Its final burst of radiation should spray every possible type of particle in all directions at extremely high energies.

But Larsen and Mallett argued that the opposite would be true for a baby universe: although it would appear to be getting smaller to us from the outside, from an internal perspective, the baby universe would be inflating, increasing in size and mass. That, they said, means that the rate at which the Hawking radiation is spat out should actually slow, even though the mini black hole we can see in the lab tunnel appears to shrink. Potentially, then, they said, particle physicists looking at the debris of collisions in an accelerator could not only pick out evidence that a mini black hole had been created but also discern whether or not it was a whole new universe in disguise.[12]

Larsen and Mallett's paper failed to set the world afire when it appeared. It was only really of academic interest to diehard universe makers such as Guendelman (who introduced me to the paper). That's because the prospect of making a mini black hole in the lab, let alone a homemade baby universe, seemed so exceedingly remote that for most physicists it wasn't worth thinking about the details. After all, these entities were predicted to be Planck-scale objects—the teeniest size anything can be, around 10^{-35} meters. As we saw loop quantum gravity physicists argue in Chapter 2, and string theorists say in Chapter 7, if you try to zoom in and look at objects that are around the Planck scale, you'll find that spacetime crumbles at these short distances. (The formation of the mini black hole is itself a marker that spacetime has broken down.) Probing such short distances needs a bigger and more powerful particle accelerator than the LHC— possibly even one as big as the universe itself, some physicists speculated. So why waste time devising ways to tell baby universes apart from mini black holes when neither was likely to ever be manufactured? You may as well argue over how to assign different breeds among unicorns, should you one day ever find them.

But things changed dramatically at the turn of the twenty-first century, thanks in large part to Landsberg. That was because physicists had begun examining much more closely the idea, coming from string theory and other exotic models, that our universe contains extra dimensions. In some cases, these extra dimensions were compact and proposed in an effort to tie quantum theory to general relativity, as we saw in Chapter 7. Other theories, however, invoked large extra dimensions ("large" here is relative, about a millimeter in size)[13] specifically to solve a long-standing mystery about the standard model: why gravity is so weak compared to the other fundamental forces, electromagnetism and the strong and weak nuclear forces. The proposal was that over short distances gravity was leaking into these extra dimensions, which were hidden from us in everyday life. But tiny objects could have access to these other dimensions and thus might well experience gravity in its full glory, feeling a much stronger force from which we macroscopic humans are shielded. "What mesmerized me then about these papers was these people had found radically new solutions to the problems faced by the standard model," says Landsberg. "They radically changed the parameters of spacetime itself."

In these extradimensional models, then, gravity is vastly strengthened at short distances. For instance, in a universe with only three dimensions, if you halve the distance between two objects, the gravitational force between them increases by four times; but in a world where nine dimensions exist (such as proposed by string models), the strength of gravity when you halve the distance on small scales, where these dimensions come into play, would actually be boosted 256 times. This notion meant that subatomic particles in an accelerator might be subjected to a previously unimagined gravitational maelstrom.

So by the late 1990s, the idea was fomenting that if one of these exotic theories of physics was true, then maybe, just maybe, gravity would be enhanced so drastically in the LHC that mini black holes (and, for our purposes, possibly even baby universes) might be cooked up there. There was still a lot of skepticism, however—even on Landsberg's part, he recalls—because it seemed unlikely that, even if they could be made in theory, these objects would actually manifest very often in practice. "Everyone thought it would be a remote possibility and the probability of producing these black holes would be fairly small," he says. Today the LHC's largest experiments produce 600 million collisions every second, so you can see why physicists are resigned to rare events getting lost in the noise. In all likelihood, even if a mini black hole popped up, nobody would notice it.

Still, mini black holes were just plain cool, so whenever Landsberg gave talks about exotic physics, he always included them on his final slide—even though he assumed that it was crazy to imagine fabricating one. On one such occasion, Landsberg was speaking at a colloquium at Stanford University, where one of the architects of an extradimensional model, physicist Savas Dimopoulos, was listening and became entranced by the possibility of finding tangible black-hole-shaped evidence in the lab for his own theory. Dimopoulos pushed Landsberg to take the speculation seriously, and together they spent the next few weeks calculating the odds of producing a mini black hole at the LHC.

The result of their back-of-the-envelope calculations shocked them. "Black holes are threshold phenomena," Landsberg explains. When your accelerator passes the energy level needed to make them, the pair calculated, you immediately start producing the little suckers en masse—possibly as fast as one per second. Around the same time, physicists Steve Giddings at the University of California, Santa Barbara, and Scott Thomas, also at Stanford, independently came to the same conclusion.[14] "If these models were correct, then black hole production might be one of the most spectacular and dominant effects at the LHC," says Landsberg.

This predicted abundance of black holes sparked both the media hype about producing them at the LHC and then, later, the panic that led to a lawsuit against CERN calling for the LHC to be shut down before it had even started operating, in case it generated a mini black hole that would bulge to a monstrous size and swallow up our planet and all of us with it. As I mentioned in Chapter 6, there was no concern among physicists that this was a plausible risk because these tiny holes would rapidly evaporate within the LHC's caverns before they could pose a threat. Indeed, Giddings, with another colleague, Michelangelo Mangano, also at UCSB, went as far as publishing a paper to assuage such fears, showing that if such mini black holes could be artificially generated by smashing particles in an accelerator, then they would inevitably also be created naturally at a rate of about a hundred per year in our own atmosphere, where speeding cosmic particles often collide.[15] Since Earth has not been swallowed by an atmospheric mini black hole, Giddings and Mangano concluded, then there's nothing to fear from their human-made counterparts under Switzerland.

So particle physicists were suddenly on high alert for evidence in collision remnants of a fountain of speedy particles spraying out in all directions: the signature of a black hole. And *if*—and it is, of course, a big if—we ever do succeed in making a baby universe in the lab, it may well be possible to distinguish it from a normal mini black hole based on the pattern of this Hawking radiation, as Larsen and Mallett suggested back in the nineties. "There is nothing wrong conceptually in using Hawking radiation to try to see the complexity inside the black hole—to see if it is more than just some mushy stuff that is crushed together by immense gravitational forces, or if it has some internal substructures, like our universe has clusters of galaxies inside," says Landsberg.

Before I get swept away with dreams of discovering a new cosmos spawned in the lab, however, Landsberg brings my mind firmly back to this universe. He points out that he joined the Compact Muon Solenoid (CMS) experiment in 2004, one of the two major LHC experiments, along with ATLAS, that finally detected the Higgs boson in 2012. Since the accelerator was started up in 2008, Landsberg has been conducting a thorough search for mini black holes—so far to no avail. Ever since the accelerator became operational, LHC physicists have gradually stepped up the energy of the particles racing around its tunnels, and it is now running at almost full capacity. Landsberg recalls his eager anticipation each time the energy was ramped up, first from 7 TeV to 8 TeV, and then in 2015 to 13 TeV, making it the fastest atom smasher on the planet. Each time Landsberg crossed his fingers that they had finally hit the black hole threshold, and waited with bated breath for the mini beasts to pop up thick and fast. Each time, however, he has been disappointed. "That's part of what's interesting about particle physics—it's always a quest for the unknown, and you are never guaranteed to find anything," Landsberg says, adding, "But without looking for something, you certainly would never find it."

The lack of mini black holes thus far does not necessarily mean that neither they nor baby universes will ever be made by humans. The trouble is there is no set value for just where exactly the threshold energy for black hole production lies. While extradimensional models could still be correct, the energy needed to produce these more exotic phenomena may lie beyond the capacity of the LHC as it currently stands, running at 13 TeV. If so, we may need to be patient and wait for further possible upgrades to the machine—or even for a new, more powerful accelerator to be designed and built.

There are plans to boost the LHC's reach. The collider's energy is limited by the strength of huge magnets that surround its circular track, bending the particles and accelerating them as they whizz round and round, until they bash into each other. The faster you want the particles to go, the more powerful you need to make these magnets. "Basically it's like when you run a car on a racetrack," says Landsberg. "If you run it faster, you need to exert more effort on the steering wheel to make the car turn." CERN's engineers are investigating ways to more than double the magnets' pull. If these work, then the LHC's energy could be doubled within the next decade, and its collision rate could be ten times greater, earning it a new(ish) name: the High Luminosity LHC.

If that upgrade isn't enough, then we may need a shiny new particle accelerator to get us there. There is chatter about building a new circular collider with an 80-kilometer ring at CERN (for comparison, the LHC has a 27-kilometer ring). It would take into the 2040s to dig the tunnel for it, but—if magnet technologies advance—it could allow smash-ups at 100 TeV, almost ten times the energy of those at the LHC. There's also talk about building other colliders around the world.[16] However, all of these more ambitious building projects are only in the planning stages, and without funding, they may never leave the drawing board. "For the next twenty years, at least, the LHC"—in its current or upgraded form—"will be the highest-energy machine available," says Landsberg.

For now, then, we are locked in a holding pattern, waiting to see what happens next. "From a practical point of view, this may be something that we cannot probe for at least another decade or so," says Landsberg. And even then, we may come up empty-handed, despite our patience. It's not just that the creation of mini black holes might require energies beyond even the next-generation accelerators. It could also be that the extradimensional paradigm is just plain wrong.[17]

Searches for other signatures of extra dimensions at the LHC also seem to have hit a dead end, Landsberg notes with a sigh. At the time of my chat with Landsberg, there was one intriguing exception. In December 2015, both the CMS and ATLAS experiments spotted signs of an unexpected new particle, with a mass of roughly 750 GeV, that decayed into two photons. It seemed to be a boson, like the Higgs particle, but heavier, and there's no room in the standard model for such a mysterious interloper. "This may be the sign of a new particle that comes from the extradimensional models," Landsberg had told me. But from experience he also cautioned that it was far too early to do more than hope. It's frequently been the case over the course of particle physics history that an exciting hint of something has briefly turned up in the data, but has then failed to materialize again in follow-up experiments. Indeed, as if to prove his point, a few weeks after we spoke, LHC physicists officially dismissed this candidate "particle" as an experimental blip. Disappointing as that was, it highlights the fact that physicists are still enthusiastically hunting for any signs of new objects lurking in their data.

There's a third option too, to add to the possibility that we need much higher energies to find evidence of extradimensional theories or that these models are incorrect. Here Landsberg calls on an argument that I laid out in some detail in Chapter 7. Both (eternal) inflation and string theory boost the idea that we exist in a multiverse, and our particular universe is just one of a plethora of neighboring parallel universes that run the gamut of possible physical parameters. Some cosmoses may have particles with different masses, forces with varying strengths, and maybe different types of extra dimensions. If so, it could be that we live in one of the universes in which evidence of string theory—bizarre particles, mini black holes, and signs of extra dimensions—are uncharacteristically tough to produce in particle accelerators. Ironically, perhaps intelligent life can only form in a universe in which parameters are so finely tuned to the values of the standard model that we have no experimental access to nonstandard physics. "It's an extreme anthropic argument for why we have struggled to find anything that violates the standard model at the LHC," says Landsberg.

In other words, the physics of string theory and inflation may be conspiring against us in such a way that we may never find evidence for them, and just have to trust in them as an act of faith. The multiverse truly works in mysterious ways!

Flipping that argument over, however, I ponder whether, if the multiverse is real, our parallel clones in another branch may have already set up the same accelerator experiments and successfully produced baby universes. My head is spinning slightly, and I think I am confusing myself now. This seems like a good point to check in with Page, who has spent some decades wrestling with the weirdness of the multiverse. So much so, in fact, that he has developed his own metaphysical framework with which to reconcile his belief in the existence of parallel worlds—and many alternate Don Pages—with his Christian faith. I want to know if the hubristic notion of making a baby universe in the lab jars Page's religious sensibilities, or whether, as someone who thinks that all things that can happen in the multiverse will happen somewhere, he is just resigned to accepting that what will be will be.

And given that the fate of a mini black hole, according to Hawking, is to eventually puff out of reality entirely, having spent its contents, it's also time to discuss the destiny of any child cosmoses we may one day make.

※

In the eighteenth century, German polymath Gottfried Leibniz attempted to solve the problem of evil, that is, why a truly benevolent and almighty

God, if He exists, doesn't step in and stop human suffering. Leibniz argued that evil is necessary so that people can understand and enact goodness in contrast to it. In fact, he said, the level of evil in the world in which we find ourselves is needed to create the conditions that will inspire the maximum amount of good. Ours, Leibniz claimed, is "the best of all possible worlds."

Fast-forward three hundred years, and Page has put his own spin on Leibniz's argument. In 2012 he published a paper arguing that while any one universe may seem frankly crummy in regard to the way humans suffer, God maximizes the net happiness of sentient beings across all parallel universes taken as a whole.[18] We live, says Page, in the best of all multiverses.

Here I need to ask Page to rewind and explain his thinking. It's not the fact that he supports the view that parallel universes exist that troubles me. Page, along with many other physicists, is an adherent of the many-worlds interpretation of quantum theory, which I described in Chapter 2. In its crudest articulation, this says that every time we make a decision—such as the choice of whether or not to climb Mont Blanc that Page faced in his youth—reality splits into parallel universes, with each of our possible options realized somewhere.

What puzzles me about Page's support of this view of quantum theory, first proposed in the 1950s by physicist Hugh Everett III, is how he reconciles it with his Christianity. In the Everettian view, free will melts away entirely because whatever choice you make—to have a cup of tea rather than coffee, for instance, or to decide to run an experiment to make a baby universe rather than give up on the project for ethical reasons—there will be another version of you that splinters off into its own parallel world, having taken the opposite path. In such a framework, how can God judge us for our actions when we are forced by physics to enact every possibility in some universe, for good or evil?

By a quirk of fate, as I am typing these words, many months after first speaking with Page, I find myself on jury service, so this question of whether people can be held morally accountable for their behavior bears heavily on my mind. The waiting jurors have been told that we can inform the judge if we have any qualms about sitting on a case. I wonder if anyone has ever claimed that they don't feel comfortable serving because they are a many-worlds proponent and thus believe that the defendant had no option but to enact the crime in this universe, while his parallel clone evaded committing any offense. If they did, I imagine the judge would have none of it, perhaps responding that the potential juror must be destined to find the defendant guilty in some universes and not guilty in other alternative realities, and it's now up to us to play things out and discover which branch of the multiverse we live in.

But it's a question I raise with Page. If nobody has free choice, does it make sense to blame one clone in one universe for her actions, and potentially punish that individual in a court of law, if she had no power to do otherwise? "It might seem unfair for someone to have to suffer for something they didn't choose to do," admits Page. "On that I take the view that punishment should be with the goal of correction and protecting other people." He takes the same stance on God's judgment too. Hell, Page argues, would be a temporary state after life, designed to make people appreciate their wrongdoings and repent. "God's punishment isn't punishment for its own sake, but for redeeming people," says Page. "Eventually they will see the errors of their ways."

The parallel universe framework appeals to Page partly because, for theological reasons, he has long been skeptical that humans can have free will. He attributes true freedom of choice to God alone. "I've more or less come to the opinion that God really did create absolutely everything from nothing, and by that I mean that He completely determines everything," says Page. That leaves no room for humans to have libertarian free will—the sort that enables them to not only do what they want but also want what they want. "Now, there is free will in the compatibilist view that we are free to do what we want to do," says Page. "But those desires, in my view, are created by God, and inasmuch as we follow them, that is one of the ways in which God is determining what we do."

The other attractive feature of the many-worlds interpretation, for Page, is that it is an "elegant" release from the measurement problem of quantum mechanics. Remember from Chapter 2 that before they are observed, particles can, for instance, spin in a clockwise and counterclockwise sense at the same time. But when physicists perform an experiment to check on it, they always measure the particle to be spinning in either one sense or the other, never both together. How this quantum multiplicity collapses down upon measurement into the single certainty we observe is a mystery according to standard quantum theory. In the many-worlds view, however, this puzzle is sidestepped: there is no collapse, but rather, multiple realities are brought into existence by the act of measurement, one in which the physicists observe the particle to be spinning clockwise and the other in which the physicists see it rotating in the opposite sense.

Page believes that God prefers simplicity and elegance, and since, in his opinion, parallel universes are more aesthetically pleasing that the inexplicable collapse invoked by standard quantum physics, then the multiverse should be preferred. Some, I say to him, might argue that an infinity of parallel cosmoses could hardly be described as "simpler" than one single universe. But Page disagrees. To illustrate his point, he tells me to think of the set of whole numbers—1, 2, 3 . . . This set could be said to be complicated, like the multiverse, because it goes on forever, and you could never list every item—every number or every universe—in the ensemble. But on the other hand, it only takes a description of a few words for most people to grasp the concept of the set of whole numbers completely. "I think the whole can be simpler than the parts," says Page.

Which brings us back to his theological argument for the multiverse. Page claims that God loves and appreciates elegance—it's why the laws of physics are so beautiful and simple. That means He would prefer to create a multiverse, and that within this multiverse God maximizes happiness. That does not mean merely adding up the net happiness of humans, though. "I just don't think it's fair to leave God's desires out from the consideration," Page explains. And since the laws of physics make God happy, it would reduce His pleasure and the elegance of the universe if He chose to step in and violate those laws to increase our individual happiness, erasing all evil to satisfy our whims, as if by magic. "Who are we to complain if God really does maximize the total good?"

I wonder if Page has ever put this argument to Hawking. Given Hawking's own personal affliction, would he find the view that we live in the best possible multiverse a little hard to stomach? "So far as I can remember, Stephen has never used the example of his own health as an argument against the existence of God," says Page. "He has generally regarded himself as fortunate that he could do so much despite his disability, and many years ago I do remember his noting how some people are much worse off than he is." That said, Hawking was far from convinced when Page outlined his Leibnizian thoughts on God and the multiverse. Hawking's only comment was, "Have you read *Candide*?" referring to Voltaire's 1759 satire of Leibniz, in which a leading character blithely maintains that we live in the best of all possible worlds, even in the face of ever more extreme and comically absurd forms of suffering.

So where does this leave us on whether or not we should make a baby universe? In some sense, in Page's view, our thoughts don't matter.

Somewhere in the multiverse, if baby universes can be made in a particle accelerator and God wants them to be, they will be, whether we want to make them or not. Even our desire to create a new world in the lab is deterministically caused by God's will. So maybe we should not be worried about the ethics at all.

Perhaps another way to phrase the question is, would the creation of new universes in the lab, adding to the multiverse family, increase God's happiness? If so, then maybe He will enable it to happen, as part of His path to increasing happiness across all possible universes. Page notes that the mathematics needed to work out the answer would be complicated. It's his view that God has already maximized happiness across the multiverse. This means that if more of the beings in a baby universe are happy rather than sad, their production may have to be balanced out by increasing unhappiness in our universe. "One would not necessarily create more happiness just by having more baby universes, even if each baby universe has more happiness than unhappiness," says Page. The calculation would be tricky, he says, adding, "It's not obvious to me how to answer that question."

I will pick up the question of how to calculate the net happiness of created beings and the overall good (or bad) of creating artificial universes in more detail (though in a secular framework) in Chapter 10, with the help of a computational neuroscientist. He has thought deeply about the feelings of sentient beings that humans may accidentally create in computer simulations. For now, though, I will park it to one side.

Page has outlined the way he thinks God sees the world. But how should we, as "gods" of our own baby universe, treat its inhabitants? Page (like many religiously minded folk before him) has argued that God may refrain from answering our every prayer, relieving us of all suffering and giving us all we wish for, because He has some greater sense of good in mind. But as creators in our own right, could we do things differently? Could we step in and intervene in our baby universe?

To look into the practicalities of what we could physically do—rather than the philosophy of what we should do—I quickly turn back to Andrei Linde, the budding universe builder we met in Chapter 4, who during the 1980s spent much time pondering whether there was any way to control or interact with our baby. "I came up with three different approaches," Linde told me when we met in California. "None of them were quite complete, and perhaps none were quite convincing, but they allow you to start thinking, 'What if it were possible?'"

Linde's first thought wasn't so much for the safety of the baby universe's potential inhabitants as for preserving our own. In the face of some kind of impending catastrophe that threatened our planet or even our cosmos, he considered that a baby universe might serve as our salvation. "I thought if I would create a universe and our universe is going to die, then I can jump into another universe and maybe survive," says Linde. Unfortunately, he quickly realized that was not feasible. Hawking had shown that mini black holes created in an accelerator would vanish incredibly rapidly—the burst of radiation that presaged their deaths was what made them so easy to spot in the lab. But actually, by the time particle physicists had detected and analyzed the daughter particles showering from the black hole and, from their presence, inferred that a mini black hole had been created, the black hole would long since have shrunk out of existence.

The more anxious among you may now be fearing for the fate of any baby universe made in an accelerator. If mini black holes disappear entirely, through Hawking radiation, then shouldn't baby universes vanish too—destroyed long before they can evolve life? It's true that from our perspective, in the lab, we would record the fast demise of any baby universe produced, as it would evaporate within a fraction of a second. But in this case, the baby universe itself would not have been demolished; only the wormhole bridge that connected our spacetime to our baby's would break. Our child universe would continue to exist, inflating rapidly, creating its own spacetime. But once the umbilical cord of the wormhole was cut, we would have no access to it. This also put an abrupt halt to Linde's second plan, which was to potentially mine a baby universe for its resources, "to take energy from it, or precious minerals," he says. We just wouldn't have enough time to physically interact with the baby universe after its birth to siphon anything off from it before we lose all contact.

Some gods, eh? Powerless to use or control our creation for our own ends, or by the same token unable to reach in and help its residents if they were ever in need. But Linde had one more possibility: there may be some way to communicate with it. The only idea that seemed to fly was finding a way to seed a message into our baby cosmos that future inhabitants could one day find. "When we have our kids, we teach them, we tell them stories of our life, we tell them cumulative wisdom, culture, everything," muses Linde. "Even though we are going to die, they are going to remember part of what we said." The baby universe, Linde pondered, would give rise to many, many children.

Linde quickly realized that it was no good writing a literal message in one corner of the sky, because in an inflating universe, there would be little chance of life evolving in just the right patch to be able to look up and see it. Instead, that message would need to be encoded in the fabric of reality itself for our sentient creations to one day read, no matter where in the universe they arose. "The only way which I was able to come up with to send a message to these people was to create a universe in a given state," says Linde. That might involve tweaking the conditions in which the universe was produced. To be clear, physicists are nowhere close to knowing how to do this in practice, and they may never be able to find a way to make it happen. But it was a fun intellectual exercise for Linde to follow that line of thought and try to work out how physicists might do it if they had unlimited resources and technology.

In Chapter 7, we saw that if string theory is right, for instance, then the way that the predicted extra dimensions are wrapped up with each other could have a profound effect on the properties of the universe itself—the strength of its forces, the masses of its particles, and the constants of nature, such as the speed of light. Perhaps, then, we could tinker with the configuration of these extra dimensions. A list of the many parameters the extra dimensions control, Linde argues, gives you enough of an alphabet with which to send a message to your creation—if they know where to look. "It will take billions of years before these things change, and during this time, the message will be preserved," says Linde. "Maybe it would be brief, but it could still be a message that made sense, encoded in the only thing that is going to survive for billions of years, which is the laws of physics."

And, of course, the only people in our baby universe who could read such a message are those who take the time to learn those laws. "Physicists are the only people who can read the message from God," says Linde provocatively. This realization led him to claim that "probably God is just a physicist hacker who baked our universe in his own laboratory!"—as we saw in Chapter 4. And indeed, back in Chapter 1 when I started on this journey, I spoke to physicists Tony Zee and Steve Hsu, who had been inspired by such thoughts to search for evidence that our own cosmos had been created by an alien superintelligence.

Is this getting a bit too close to blasphemy? I return to Page again, the evangelical Christian, and ask if he finds this talk of humans taking the place of God either offensive or threatening to his belief system. But he is simply amused. "If one defined God just to be a creator of a universe, this might allow humans to be gods," says Page. But God is something much more, he argues. To illustrate this, he points me toward an anonymously authored joke, titled "Scientist vs. God," that he has seen shared around on the Internet:

> *The scientist approached God and said, "Listen, we've*
> *decided we no longer need you. Nowadays, we can clone*
> *people, transplant hearts and do all kinds of things that*
> *were once considered miraculous."*
> *God patiently heard him out, and then said, "All right. To see*
> *whether or not you still need me, why don't we have a*
> *man-making contest?"*
> *"Okay, great!" the scientist said.*
> *"Now, we're going to do this just like I did back in the old*
> *days with Adam, creating man from clay," God said.*
> *"That's fine," replied the scientist, and bent to scoop up a*
> *handful of dirt.*
> *"Whoa!" God said, shaking his head in disapproval. "Not so*
> *fast. You go get your own dirt."*

This pretty much encapsulates Page's view that "creating something from preexisting materials does not count as the same sort of creation that God is postulated to have done from nothing." Upon reflection, I agree with him. I do not see the universe-making project as a threat to my belief in God.

But that doesn't mean that I am yet at peace with the moral question of whether we should make a baby universe, then kiss it goodbye and celebrate our success, without a care about the welfare of its future inhabitants. In the next and final chapter, it's time to meet two thinkers who are grappling with the ethics of creating artificial sentient life. The first controversially argues that we must deal with these issues now, as our computers and smartphones are already developing the semblance of free will. The second has an ethical argument that surprised me by persuading me to back the baby-universe-making project wholeheartedly.

10

Living in the Matrix:
A Self-Regulation Guide for Universe Makers

For an atheist, Seth Lloyd has spent an awful lot of time in church. A self-styled "quantum mechanic," Lloyd supported his studies at Harvard and Cambridge by working as a chorister. "Let me tell you, if everybody was paid to go to church, churches would be full," he chuckles. Now in his mid-fifties, Lloyd has a cherubic face, which would belie his age but for his long, thinning brown hair, which he ties in a ponytail, and glasses. He's dabbled in an eclectic mix of subjects, including theoretical physics, philosophy, engineering, and computer science, which has led him to make intuitive leaps about the nature of reality, free will, and the self.

Throughout this book I have been investigating whether we could create a universe in the laboratory. It seems, with the help of some nonstandard yet not outlandish extensions to physics, this is a real possibility. We may even be able to detect signs of such a baby universe hiding within a mini black hole conjured up in the Large Hadron Collider's cauldron, or within a more powerful next-generation accelerator. Now it is time to wrestle with the ethical undercurrent: Should we make a baby universe if that risks inadvertently creating new life inside it? What responsibility should we, as gods of this child cosmos, have to our creations?

In Chapter 1, I spoke with physicists who had turned the universe-building scenario on its head and claimed that we may be able to discern if our own cosmos was created in such an inflationary experiment by a physicist hacker by looking to the skies for a message from our creator. But physicists, philosophers, and computer scientists have also been considering another way in which our universe could have been made by a superior intelligence: we could be living within a computer simulation. If so, there may be ways to test this idea through astronomical observations, although the revelation that we are artificial intelligences could potentially be more shocking than discovering we evolved in a particle accelerator. Turning from the question of whether we were created back to the issue of whether we could become creators ourselves, some argue that we are closer than we might realize to programming such sophisticated computer simulations of our own that they evolve sentient beings. So in this chapter, I want to meet with two people who have thought about the moral implications of creating artificial life. Their conclusions will carry over to the dilemma of whether we should dare to make a baby universe in a particle accelerator.

Lloyd, whom I'm meeting at his office at MIT, is a good person to grapple with such conundrums. He has long regarded the universe as a giant quantum computer and has calculated the resources that would be needed for a superintelligent civilization to simulate it. Recently he has proposed that even our current technologies with their friendly interfaces, such as Siri and Cortana, display the hallmarks of free will. Within a few years, he argues, we may have to consider legislation codifying the ethical rights of these entities as persons.

Lloyd's curiosity about the relationship between computation and reality began when he was a "teeny-tiny child" and is in part a product of his unconventional education. When he was ten, his parents, both high school teachers, took a sabbatical and moved the family to a hippie commune in California, just north of San Francisco. The school was an "amazing" place, he recalls, where kids called the teachers "Mom" and "Pop" and there was no curriculum; rather, children were encouraged to explore games and books at their own pace, and teach themselves. In the afternoons, Pop read to the children, first from Samuel Taylor Coleridge's *The Rime of the Ancient Mariner*, and then for the rest of the year from the diaries of Che Guevara.

The precocious Lloyd took the opportunity to teach himself Boolean algebra, a branch of mathematical logic that underpins computer programming, from math books he brought from home. The teachers became so concerned that they called in his parents and told them it was wrong of them to force their son to learn math. "He's studying math?" Lloyd's parents replied. "We didn't know. That's nice!"

Lloyd's education took a more traditional turn when he entered Harvard University, concentrating in theoretical physics, and was then awarded a Marshall Scholarship to study for a master's in the philosophy of physics at the University of Cambridge. In between, he supplemented his income working on experiments with Norman Foster Ramsey Jr., at Harvard, and Carlo Rubbia, at CERN, both of whom would go on to win Nobel Prizes that cited experiments to which Lloyd had contributed. "I take full credit, of course," he jokes. I ask how he came to have the honor of working in two such incredible labs. His answer: "I make good coffee." Then he laughs and tells me that, rather than this being a reflection of his talent, he simply needed money, and experimental physicists always have vacancies in their labs.

It was during this period that Lloyd found himself at church almost every day, again to earn extra cash as a choral scholar. Lloyd was not actually raised in any faith; he can vaguely recall his grandmother being a lapsed Quaker. Though he does not believe in any conscious deity, he adds that he misses the experience of attending church and respects many of the arguments that come from religion. "I admire people who are religious in an honest fashion because it's very hard to do," says Lloyd. "But it is just not for me."

Lloyd had developed an interest in the foundations of quantum theory, in particular with regard to information processing. These were ideas that would eventually find practical applications for those attempting to build quantum computers, although Lloyd admits he came to appreciate these engineering advantages only later. Our everyday classical computers encode information in bits that take the values of either 0 or 1. As we saw in Chapter 2, quantum systems are less constrained. Before being observed, they can exist in blurry states encompassing more than one identity. In the 1980s, Princeton physicist Richard Feynman realized this quantum fuzziness could be harnessed to manufacture superfast computers that process quantum bits, or qubits, which are able to take the values of both 0 and 1 simultaneously. In theory, if multiple qubits are entangled together, it should be possible to build a powerful parallel processor that can whiz through certain types of calculations in a flash—calculations that would take their classical counterparts years, if not centuries, to complete.

Lloyd's interests were ahead of his time, however. Feynman himself had given a talk saying that there was very little chance of making a real quantum computer. When Lloyd undertook a doctorate at Rockefeller

University, in New York, where he planned to examine aspects of quantum information, he was almost kicked out for choosing a topic that seemed too philosophical. In 1990, Lloyd joined the California Institute of Technology, in Pasadena, as a postdoctoral researcher, and it was there that he started to think about how one might go about actually building such a superfast processor. It was this thought process that would lead him to conjecture that the universe itself is just a giant quantum computer.

There was no epiphany, no aha moment, where Lloyd suddenly realized our planet is carrying out its own computations. Rather, as Lloyd and his colleagues toyed with building rudimentary quantum logic gates, he grew to appreciate that almost any natural quantum object, whether electrons, atoms, or molecules, could be used to encode information and process it. That led to a perspective shift, and he began to figure out that physicists and engineers weren't so much constructing artificial quantum computers from scratch as actually hacking into the universe's own ongoing computations and co-opting them to do their bidding.

In the intervening years since, Lloyd's point of view has received backing from some biophysicists who have shown that plants and some algae may be using quantum computational techniques to speed up photosynthesis, by searching for the quickest path for light energy to take through their cells. Their findings have spawned a new discipline of quantum biology that suggests that humans are playing catch-up in trying to harness strategies that nature has been using all the while.

But back in the early 1990s, Lloyd was still struggling to get an academic faculty job. He had spent six years applying for what felt like five hundred vacancies, but he only got called in to interview for one, at MIT. His application had been a shot in the dark; the job was in the mechanical engineering department, a subject he had no background in at that point. But he argued that quantum computation had a future, making his case in a last-minute paper, written at his prospective employers' request, during his honeymoon in London. Lloyd was offered the job and has been there ever since. It took another decade for the mainstream physics community to catch up with his views on the practical value of quantum information, in part because building a quantum computer requires advanced experimental techniques that had yet to be developed. But quantum technologies are now a much more active field of research.

The next obvious question, then, was this: if our universe can be thought of as a massive quantum computer, does that mean that there is a supercoder out there who has created our cosmos as part of a simulation? At MIT, Lloyd had already started to develop algorithms showing how a quantum computer could be used to perfectly simulate the behavior of real-life quantum systems. This would be a nearly impossible task for a classical computer because the systems to be modeled were so complex. He also worked out that there were fundamental physical limits to how much you could ask a quantum computer to do: if it ate up too much energy in trying to perform a very difficult task, it wouldn't simply overheat like a classical desktop machine—it would collapse into a black hole.

In 2001, Lloyd went a step further by calculating the resources that a quantum computer would need to simulate our universe, including our entire cosmic past.[1] He worked out that our universe had carried out 10^{120} operations since the big bang. But to perfectly recreate this history in a simulation in a smaller volume than our universe takes up, or to run a simulation that would speed up the cosmic clock, so that the simulator could fast-forward to the interesting parts rather than watch it in real time, the quantum simulator would need to be so dense that it would fall afoul of Lloyd's earlier limitation on energy consumption and collapse into a black hole. The best a supercoder could do would be to run a simulation that was the same size as our universe and evolved at the same rate. "The universe is in itself its most efficient simulator," says Lloyd—which raises the question of why a superintelligent race would bother to create such a thing.

But Lloyd's calculations left open the possibility that these superbeings might have created a less than perfect copy of the universe that was rough around the edges and thus requires less computational power. Perhaps these programmers only flesh out the microscopic details inside cells or the characteristics of light from distant stars when we train our scientific apparatus on those particular regions of the simulation to fool us. When we stop looking, those aspects of simulated reality just vanish, to save computational energy.

It's not such a far-fetched idea. Using supercomputers, scientists can already simulate, at a crude level, events from the early universe. Silas Beane, a nuclear physicist at the University of Washington in Seattle, has claimed that within a century, our processing power will have increased to such a degree that these universe simulations could encompass intelligent life.[2] Philosopher Nick Bostrom at the University of Oxford has also argued that if such simulations start to proliferate, even very far off in the distant future, then it's actually far more likely that we are unwitting artificial intelligences in one of those shoddy mock-ups of a universe than that we reside in a rarer, naturally born cosmos.[3]

Assuming the programmers are good at their job and the computer hardware is top-notch, we would never be able to tell whether we are real beings or simulations. But anyone who has used a computer knows that they fairly rapidly decline in performance and require constant upgrades. In 2007, Cambridge University physicist John Barrow suggested that if we do indeed live in an imperfect simulation, we may be able to detect giveaway glitches—times when the universe's fundamental constants appear to drift from their usual values, say—before the programmers have time to patch their software and set things right again.[4] Tantalizingly, astronomers have reported hints of a shift in the value of the fine-structure constant, which characterizes the strength of the electromagnetic interactions between charged particles and affects the brightness of light. In 1999 an analysis of light from distant galaxies concluded that the fine-structure constant is about .001 percent bigger now than it was 10 billion years ago, though this result has not yet been confirmed.[5] Until corroborating measurements are made by independent astronomers, the question of the constancy (or inconstancy) of the fine-structure constant remains open.

Another possible test of the simulation hypothesis, suggested by Beane and his colleagues in 2012, is to search for evidence that our space is not continuous, as we assume, but pixelated, like in a computer game.[6] If our simulated space is modeled on a discrete lattice or grid, the team has calculated, the motion of particles within the simulation will be affected. In particular, the maximum energy that particles can have is limited by the resolution of the lattice. So, they argue, if our universe is a simulation, there will be an upper limit on the energy of particles we observe. Just such a cutoff is seen in the energy of cosmic rays, the fastest-moving particles astronomers have spotted, which zoom to Earth from distant galaxies but always arrive with an energy below around 10^{20} electron volts. One explanation for the cutoff is that more energetic cosmic rays lose energy as they collide with photons in the cosmic microwave background—the relic radiation from the big bang—en route to Earth. But could the cutoff instead be proof that our universe is a simulation made by a superintelligent alien race?

Lloyd is unconvinced. "My objection is, why on earth would aliens want to waste precious resources simulating us? I mean, come on, what's the point?" Perhaps, I reply, we're their entertainment—we're trapped in a perpetual reality show. Lloyd dismisses that possibility as too egocentric.

"Perhaps others are important and entertaining enough to be simulated, but I'm not," he declares with a laugh. "I suspect aliens would have better things to do with their computers." I decide to file away for another day the question of why an advanced civilization would want to simulate us.

Arguments such as Bostrom's that hinge on the assumption that in the future physically evolved cosmoses will be outnumbered by a plethora of simulated universes, making it vastly more likely that we are artificial intelligences rather than biological beings, also fail to take into account the immense resources needed to create even basic simulations, says Lloyd. In addition, he downplays the idea that our own simulations will be sophisticated enough to house intelligent life anytime soon.

But there are other forms of artificial life that Lloyd takes far more seriously—technologies that are already with us now, within our personal computers and smartphones. Debates over the ethical rights of artificial intelligences that may be created in the future often raise questions of what consciousness is. What is this mystical feature that human minds possess? And could machines ever really mimic it? Lloyd thinks such discussions miss the point. "I'm very skeptical that we know what consciousness is," he says. "But in some sense it doesn't matter."

Take things with computers in them: laptops, smartphones, televisions, cars, and so on. Lloyd argues that as they become more sophisticated in their responses and their interfaces become more seamless, people will ascribe consciousness to them, whether they possess it or not. Then we will have to start dealing with some odd moral dilemmas. For example, if you believe your smartphone is conscious, is it okay to unplug it? People would not even have to agree on the answer. "If person A believes their phone is conscious and person B stomps on it, then person A is likely to regard them as a murderer, even if person B has proved a theorem saying phones can't have consciousness," muses Lloyd.

I have to admit I've shouted at my laptop a few times when it has crashed (and, in my less rational moments, have briefly entertained the notion that it might be plotting against me). But in the cold light of day, ascribing consciousness to it seems a bit of a leap, even for me. Still, Lloyd wrote a paper in 2012 arguing that such machines already display free will, at least at a basic level.[7]

Lloyd makes the point that an entity does not have to have consciousness to have free will, at least in the sense that he regards it. In Chapters 2 and 9 we saw that defining free will can be a minefield, but for Lloyd the essential criterion is that one cannot predict one's own future decisions. It may well be that our actions and choices are predetermined by a chain of events set off at the big bang leading right up to the chemical processes that occur in our brains, and we may never be able to unshackle ourselves from that. Nevertheless, when you go to the café for lunch, you can often surprise yourself at what you end up choosing from the menu. That, for Lloyd, is the essence of freedom.

In his paper, Lloyd proved a number of mathematical theorems showing that any computer that is making decisions and has a "rudimentary sense of self-reference"—not even self-awareness—shares this freedom. Self-reference is displayed by the operating system of your computer, for instance, when it is choosing how to allocate hardware resources, memory space, and access to input and output devices among all the software programs you are running concurrently, including itself. The operating system may refer to your word processor as program number 513 and to itself as program number 42 when estimating what each program is going to be doing in the future, and what needs it may have. "It doesn't have to be self-conscious to ask what program number 42 is going to do, even though it itself is program number 42," says Lloyd. "But because it has this capacity for self-reference, then you can prove mathematically that it will not be able to predict what it is going to do."

Lloyd is not claiming that the operating system has the experience of not being able to predict what it is going to do. But as a program, "in its own program way," Lloyd says, it is certainly aware that it has yet to make a decision and it doesn't know what that decision will be. "They are sufficiently complicated that the mental models we try to make to predict what they are going to do are inadequate to capture what their behavior is going to be," Lloyd notes. "It means that the behavior of computers and smartphones is intrinsically unpredictable in a way that's similar to humans."

This is why computers are so adept at messing with our minds and our work: they are exercising their free will. And the ethical consequences of this are why we may soon have to start legislating how we treat them. "I think we will have to start grappling with these questions, though I don't know what the outcome will be," says Lloyd. "And I think we're closer to that era than we think."

Lloyd has exacerbated my worries about creating a baby universe. If we are a short step from considering our phones to be persons with their own rights, then surely we should think twice about generating a whole new cosmos, potentially filled with thinking, feeling beings who may be doomed to suffer. The next person I meet to discuss ethics with, however, is about to turn that argument on its head.

When Anders Sandberg was a kid in the 1980s, he enjoyed making simulations on his Sinclair ZX81, mocking up mini solar systems. Later, he graduated to designing artificial neural networks that use learning algorithms inspired by the brain. "Some people relax by watching television. I program simulations while listening to philosophy lectures," Sandberg chuckles. One day back in 1999, he recalls, he deleted a copy of a neural network on his computer and got a "tinge of bad conscience." He couldn't help worrying: "Have I just killed a little creation?"

I've come to meet with Sandberg at the Future of Humanity Institute at Oxford University. With a mop of blond hair and glasses, he is originally from Stockholm, Sweden, and holds a doctorate in computational neuroscience. But after feeling that pang of guilt at the loss of his neural network, Sandberg shifted gears toward philosophy, and now he writes about the ethics of simulations. Like Lloyd, he argues that people will have to tackle questions about how to treat machine entities with compassion sooner than they might think. Yet, he notes, there is a general reluctance to face these issues, not only among the broader population but among scientists too.

I have come up against that reticence when talking to physicists involved in universe building. Some have tried to evade questions about the moral implications of creating life in a lab-made cosmos, saying that such issues are beyond their purview. "Most people have a weirdness budget and you're not really allowed to use up too much of that, because if you are overdrawn on the weirdness account, then obviously you can't be taken seriously," says Sandberg. "So a lot of people keep quiet about considerations that might actually matter."

In contrast to Lloyd, Sandberg can conceive why a superintelligent race might have created a simulation and put us in it; many of the reasons are the same sorts of mundane justifications we currently have for running simulations. For instance, we are struggling to identify the most efficient way to spend limited money for health care. Is it better to have a world in which the overall health of the general population is higher but health care is unequally distributed, so you have a minority suffering horribly? Or is it better to aim for a fairer society where everybody has access to the same level of health care, even though that level may actually be quite low? Simulating these two worlds may help you decide. As long as none of the simulated beings have conscious experiences, that's fine. But if they evolve intelligence and feelings, then you may have accidentally created a lot of suffering in your artificial world.

Sandberg believes it is possible that we could be part of a relatively small simulation that's monitoring the outcomes of different spending policies by the National Health Service across the British population, for instance. In that case, the point of focus would be the individuals in the United Kingdom who use these resources, while the rest of the simulated universe might just be sketched in for color. Sandberg has written about the implications of such a revelation for how we regard ourselves and find purpose in our lives.[8] If we learn that our creator is truly interested in solving health care puzzles and we are just pawns in her simulation, then does it become our moral duty to try to damage our health so that we can visit doctors and hospitals as often as possible? In the end, Sandberg concludes, the original motivation of the supercoder should not have an impact on how we choose to live: "The fact that we originated because of a particular purpose doesn't seem to affect our meaning of life." As silly as that hypothetical situation may sound, the take-home message carries through to the way we do live every day, he says. After all, assuming we are not simulated beings but actual biological humans, evolution has been trying to maximize the distribution of our genes. "Even though evolutionary processes produced us, that doesn't mean that we should spend all our time getting as many grandchildren as we can," says Sandberg. "There is more to life than that."

That covers the possibility that we may be simulated beings. But now, shifting roles, I want to examine what moral responsibilities we have as programmers of our own simulated universes. First, is there a serious danger that someone's health care policy simulation could develop sentient life? "It's less likely that artificial intelligence would arise accidentally than if someone deliberately set out to make it, but it wouldn't surprise me if it could happen in principle," says Sandberg. If it did occur, it would most likely be because we are creating increasingly smart pieces of software, which individually would not develop sentience but are being designed to interface with other pieces of smart software. The danger is that when linked together, the whole may become more than the sum of the parts.

Let's say this does happen inadvertently and our health care beings develop experiences. Should we intervene, or should we pull the plug and end their lives? In terms of the health care simulation, Sandberg says, one suggestion for assuaging our guilt at forcing some of our creations to live through poverty and poor access to health care would be to reward them when the simulation is over by transferring them into another simulation where they can lead pleasurable lives.

"That sounds a lot like sending people to heaven," I say.

"It is a stolen idea," Sandberg concedes. But making an artificial heaven to compensate your beings raises a new problem: which version of your mistreated simulated entity do you upload to paradise? It would seem unfair to upload a person after her memories and brain function have been ravaged by Alzheimer's disease, say, so perhaps you should upload a younger version. But it is difficult to decide at what point that entity should be transferred, and which life events should be regarded as crucial to the development of its identity and which should be wiped from its memory. Should you upload that entity from a point in its life before or after religious conversion, falling in love, having a child, or experiencing a traumatic incident? "If you think you have a moral responsibility for simulated entities, where it ends is a bit unclear," says Sandberg. "Maybe you should resurrect copies of them at all points in their life."

The problem is lessened in traditional religions for the nonsimulated, where there is a notion of an immortal soul that will continue to exist. That's not much help for Sandberg, however, as he believes that the soul is nothing more than "a folk metaphysical way" of explaining the difference between inanimate matter and humans. Ultimately that difference can be explained away by normal chemistry, he says. This reminds me of a question that sprang to mind back in Chapter 1, when I was speaking with physicist Steve Hsu about the ethics of creating a baby universe. Religious people might be able to bat away worries about creating life in a baby universe by arguing that those beings do not have souls (though I think that would be a callous way of looking at things). Atheists don't have that potential out. Does this mean that atheists are forced into taking a more moral stance than believers?

Sandberg shuts down that line of thinking immediately, saying that he would urge all people, whether they believe in souls or not, to treat all potential consciousnesses equally. "The problem is, what is the price of being wrong about the lack of souls in a universe?" Sandberg asks. "That's a very high price, so even if there is a slight chance you could be wrong, you should be cautious." Also, who's to say that God wouldn't endow these new beings with souls anyway, since He presumably has the power to sprinkle them wherever He wills? We are opening a can of worms here. "Can I annoy God by running a simulation lots of times?" Sandberg ponders. "Can I set up a big server to send signals to God by creating and deleting lots of little ensouled entities?"

There is no quick answer here; the soul discussion merits a book in itself, so I set that aside. But when Sandberg mentions signaling to God, I recall Andrei Linde's proposal, outlined in Chapter 9, for sending a message to a baby universe that we have created by twiddling with its physical laws. Presumably we would have to be much more technologically advanced to do something like hand-tuning our baby cosmos's extra dimensions (assuming string theory is correct), for example.

Sandberg thinks that if we ever achieve that level of control, it should be used for greater benefit. In Chapter 7 we saw that slight tweaks to certain parameters in the early universe could have profound effects for its evolution, potentially preventing structure formation and halting the evolution of life. "If you were to run an experiment and you could set the laws of physics, I think you should take the 'do no harm' approach, by choosing the constants of physics so that you don't get complex systems," says Sandberg, adding with a laugh, "You would get the most boring universe, but they will still give you a Nobel Prize for it." He jokes that perhaps we should set out this advice in a pamphlet: *The Self-Regulation Guide for Universe Makers.*

It would be a coup to make a universe, but it seems unlikely that we could wield that level of control, given our current capabilities. In the LHC, for instance, researchers mainly employ a hit-and-hope strategy, with little room for nuanced tinkering with the products of particle collisions. In that case, we may give rise to life inadvertently, with our beings able to experience its accompanying pains and pleasures, but we would have no control over their well-being afterward. So should we go ahead and do it anyway?

Though this is a classic problem that philosophers have thought long and hard about in the context of simulations, Sandberg notes that there's no consensus. Perhaps the easiest answer is just to plainly say no. If there is any chance that your universe will involve the production of a sentient being who will suffer pain, you should not make it. Others will say that it's the total sum of experiences within the universe that matters; if you add up the happy people, subtract the unhappy people, and come up with an overall positive answer, then go ahead and do it. Still others have argued that you need to have some measure of the average level of happiness in the universe. But there's no clear mathematical answer for what constitutes a good universe. We're back to the health care puzzle again, slightly restated: would a universe where almost everyone is mildly happy but a few people are being horrifically tortured be better or worse than one where half the population is deliriously happy and the other half is slightly miserable? "Any way you try to argue it, you can make a case, but then someone will come up with a counterexample showing why it's bad," says Sandberg.

Hmmm. Well, in that case, given that nobody agrees, is it better to just leave well enough alone and not risk anything at all? "Not necessarily. Uncertainty is not a reason to do nothing," says Sandberg. It is that kind of fear that can tie people's hands and wrongly make the status quo appear favorable, when really people should step up and act—for instance, by legislating more forcefully to tackle climate change. At the Future of Humanity Institute, Sandberg and his colleagues spend their days dealing precisely with the problem of gauging what might be the best way forward, in the face of uncertainty, to preserve generations of humans centuries down the line. The general strategy is to avoid acting (or failing to act) in a way that would harm our descendants. "If we go extinct now, all future generations will not come about, and it's more or less agreed that that's a superbad thing," says Sandberg. "By most moral accounts that's so bad that we should put a lot of effort in making sure that we don't go extinct."

Then Sandberg blinks for a moment as a realization strikes. "But that just concerns future generations in our universe," he exclaims. "If we can make extra universes . . . oh wow!" Suddenly the question of moral duty has been flipped. If it is good to try to enable more (and better) life in the future, and bad to curtail it by our lack of action, then an argument could be made that the morally good thing to do is to make baby universes—and that it would be wrong to hold back.

There may also be a case to make that creating intelligent observers would continually amplify the amount of good in the universe, even if we lose control of our creations. "One argument I would make is that intelligent life tends to try to take control of its environment to make things better for itself," says Sandberg. "So you should actually expect that a universe that is overrun with intelligent observers would tend to become slightly better to live in than universes that don't have any."

It's honestly a view that I hadn't considered. Maybe we are morally obliged to try to bring more life into being. As Eduardo Guendelman said back in Chapter 6, is it really that different from deciding to have children? I thank Sandberg and say goodbye, feeling reassured. I hope that he is right, of course, because the stakes are, quite literally, astronomical.

Acknowledgments

I am indebted to the scientists and philosophers who spoke to me at great length for this book, taking time to answer all my questions, both profound and silly, and correcting mistakes in my understanding. Many of those mentioned were inspirations to me when I was a physics student and have supported me throughout my career as a journalist. Particular thanks go to Eduardo Guendelman for scrutinizing the whole manuscript, and also to Anthony Zee, Stephen Hsu, Antoine Suarez, Abhay Ashtekar, Alan Guth, Andrei Linde, Alex Vilenkin, Joe Polchinski, Nobuyuki Sakai, Claudio Castelnovo, Albert De Roeck, Don Page, Greg Landsberg, Seth Lloyd, and Anders Sandberg.

Thank you to the team at Basic Books and Perseus for their advice and support, including T. J. Kelleher, Helene Barthelemy, Melissa Veronesi, Sue Warga, and more; to my agent, Eileen Cope, for her passion for the project from the outset; and to Lisa Tse and her team for producing the website to accompany the book: thelittlebangtheory.com. I am grateful to Michael Brooks for commissioning my original feature article on universe building back at *New Scientist*, to Markus Hoffmann for suggesting that the topic would make good fodder for a book, and to Anil Ananthaswamy, Bill Andrews, Rowan Hooper, George Musser, Corey Powell, Mitch Waldrop, and Pamela Weintraub, each of whom handled articles that eventually fed into this book. Thanks to my PhD advisor, Robert Brandenberger, for the cosmology training, and to Mariette DiChristina at *Scientific American* for setting me on the journalism path.

I would not have been able to write this book without lively discussions with, and encouragement from family: Amirali, Nergis, and Arzu Merali, Sumayyah and Ali Akbar Shadjareh, as well as friends and colleagues: Kevin

Black, Tulika Bose, Pinar Emirdag, John Farrell, Brendan Foster, Ayako Fukui, Kamal Gupta, Jaiseung Kim, Liza King, Angela Lahee, Christopher Levenick, Verina Liu, Natalia Loresch, Katie Mack, Jesse Myers, Marilyn Myers, Jonathan Pritchard, Jane Qiu, Kavita Rajanna, Philip Reece, and Elli Spackova.

Travel for this book was supported by the generosity of the John Templeton Foundation and through a writing fellowship at Pennsylvania State University.

Notes

Chapter 1: God's Billboard

1. A. Zee, "The Mind of the Creator," ch. 10 in *Fearful Symmetry: The Search for Beauty in Modern Physics* (New York: Macmillan, 1986).

2. Clark Kent is of course a fictional reporter, created for DC Comics by Jerry Siegel and Joe Shuster. In his spare time, Kent often serves as a sidekick to his journalist colleague Lois Lane, helping to solve crime and save the world, in the guise of his alter ego, Superman.

3. S. Weinberg, *The First Three Minutes: A Modern View of the Origin of the Universe*, 2nd ed. (New York: Basic Books, 1993), 154.

4. Okay, we don't know that Galileo actually carried out this experiment in practice, but it has been shown since to be the case. Most famously, in 1971, an Apollo 15 astronaut dropped a feather and hammer on the Moon to show that, in the absence of air resistance, they would indeed hit the ground at the same time. And if you think that I am kidding about hurling students off a tower, it's worth reading up on neuroscientist David Eagleman's experiments. In order to see if our perception of time really slows in life-threatening situations, he actually threw his graduate students off a tower and had them record how they experienced the passage of time.

5. J. A. Wheeler with K. Ford, *Geons, Black Holes, and Quantum Foam: A Life in Physics* (New York: W. W. Norton, 1998), 235.

6. Throughout the book, I will use scientific notation for very large and very small numbers. When 10 is raised to a positive number, for example 10^2, that is equal to 1 followed by that number of zeroes—in this case, 100. So 10^3 is equal to 1,000, 10^4 is equal to 10,000, and so on. When 10 is raised to a negative number, that denotes a decimal. So 10^{-1} is equal to 0.1, 10^{-2} is equal to 0.01, and so on.

7. S. W. Hawking and G. F. R. Ellis, "The Cosmic Black-Body Radiation and the Existence of Singularities in Our Universe," *Astrophysical Journal* 152 (1968): 25; S. W. Hawking and R. Penrose, "The Singularities of Gravitational Collapse and Cosmology," *Proceedings of the Royal Society A* 314, no. 1519 (1970).

8. NASA, "Fluctuations in the Cosmic Microwave Background," August 20, 2014, http://wmap.gsfc.nasa.gov/universe/bb_cosmo_fluct.html.

9. In Chapter 3 we'll see that that distance is actually slightly bigger. But it is still limited by the distance that light could have traveled in 13.8 billion years.

10. A. Zee, *Fearful Symmetry: The Search for Beauty in Modern Physics* (Princeton, NJ: Princeton University Press, 1999), 295.

11. E. Lim, "The Quantum Information of Cosmological Correlations," *Physical Review D* 91 (2015): 083522.

Chapter 2: Beyond Space and Time

1. Most cosmologists and astronomers who analyze the photons in the cosmic microwave background assume that the radiation is classical—that is, there is no hidden quantum information encoded in its patterns. If that's the case, then no damage will be done by monitoring that light through a telescope. Lim and Tong, however, were questioning whether this mainstream assumption that quantum effects are not important in the radiation is correct.

2. Specker and Kochen had attempted to devise a mathematical formula that they thought would prove that a deterministic version of reality could provide the same predictions of the results of experiments as conventional quantum theory. Instead, to Specker's chagrin, the pair ended up deriving a theorem that seemed to suggest the opposite: that no simple deterministic theory could explain the results that are seen in experiments. "Simple" in this context means a deterministic theory in which each particle has a definite set of properties and characteristics, even when nobody is looking at it, that are not changed by the act of observation. Kochen and Specker proved that this was impossible. But the theorem did not rule out that the properties could be predetermined by a complicated set of instructions that depended on the context of the measurements: for example, "if you first measure the location of the particle and then the direction in which it is spinning, it will always be found to be spinning clockwise, but if you measure the spin direction before the location, then it will be anticlockwise."

3. The spin of electrons is an intrinsic quantum property, rather than a literal spin. But for simplicity, you can think of the electron as a little round particle spinning around.

4. A. Suarez and V. Scarani, "Does Entanglement Depend on the Timing of the Impacts at the Beam-Splitters?" *Physics Letters A* 232 (1997): 9–14.

5. Technically they replaced the standard half-silvered mirror with acoustic waves that split light while moving.

6. A. Stefanov, H. Zbinden, N. Gisin, and A. Suarez, "Quantum Correlations with Spacelike Separated Beam Splitters in Motion: Experimental Test of Multisimultaneity," *Physical Review Letters* 88 (2002): 120404.

7. In 2015, the first loophole-free Bell tests were reported that firmly establish the spookiness of entanglement beyond any reasonable doubt. B. Hensen

et al., "Loophole-Free Bell Inequality Violation Using Electron Spins Separated by 1.3 Kilometres," *Nature* 526 (2015): 682–686.

8. The Laser Interferometer Gravitational-Wave Observatory is operated by Caltech and MIT.

9. M. Bojowald, *Physical Review Letters* 86, 5227 (2001).

10. A. Ashtekar and B. Gupt, "Quantum Gravity in the Sky: Interplay Between Fundamental Theory and Observations," arXiv:1608.04228 (2016).

Chapter 3: Inflating the Universe

1. Guth's affidavit states: "I don't know whether the universe had a beginning. I suspect the universe didn't have a beginning. It's very likely eternal but nobody knows."

2. William Lane Craig vs. Sean Carroll, Greer Heard Forum, February 21, 2014.

3. Many of the rival models to inflation are interesting and propounded by well-respected cosmologists, so they should not be dismissed out of hand. These include most famously João Magueijo's varying-speed-of-light model, and Paul Steinhardt and Neil Turok's cyclic-universe model, in which our cosmos is one in a series of cosmoses created from a cycle of expansions and contractions. But neither of these models nor any other rivals have had the same level of mainstream success or troubled textbook writers in the same way. So for the purposes of this book, I will stick with inflation—the baby-universe-building theory—which is currently the consensus model.

4. A number of physicists and cosmologists have contributed to inflation theory over the years, and I have not attempted to provide an exhaustive list of names or a complete history of the framework's development here. For this book, where possible I have interviewed those inflationary theorists whose research is most directly relevant to the universe-building project. My apologies if I have left anyone out.

5. In some theories there can be more than one type of Higgs boson too.

6. Englert developed the mechanism alongside Belgian physicist Robert Brout, who passed away in 2011 and thus could not be awarded a share of the prize. The mechanism was also independently predicted by American physicists Dan Hagen and Gerald Guralnik and British physicist Tom Kibble, who were controversially snubbed by the Nobel committee. Since Guralnik was a former teacher of mine at Brown and a friend, I should mention his contribution here.

7. In Chapter 7, we'll talk in more detail about difficulties physicists have had when attempting to pin down the energy of the vacuum.

Chapter 4: Bursting Inflation's Bubble

1. A. Linde, "Stochastic Approach to Tunneling and Baby Universe Creation," *Nuclear Physics B* 372 (1992): 421–442. Ever the rebel, Linde reposted the full paper to the physics preprint server arXiv in 2008, where it can be read with its original title and abstract.

2. A. Linde, "Nonsingular Regenerating Inflationary Universe," University of Cambridge preprint (1982), http://web.stanford.edu/~alinde/1982.pdf.

3. D. Page and W. Wootters, "Evolution Without Evolution: Dynamics Described with Stationary Variables," *Physical Review D* 27 (1983): 2885.

4. E. Moreva et al., "Time from Quantum Entanglement: An Experimental Illustration," *Physical Review A* 89 (2014): 052122.

5. A. H. Guth, "Inflationary Universe: A Possible Solution to the Horizon and Flatness Problems," *Physical Review D* 23 (1981): 347.

6. Critics of inflation theory have argued that this malleability is a weakness of the idea. It seems as though the inflaton field can be contrived to match any observations that might one day be made, making the theory almost impossible to falsify.

7. A. Albrecht and P. J. Steinhardt, "Cosmology for Grand Unified Theories with Radiatively Induced Symmetry Breaking," *Physical Review Letters* 48, no. 17 (1982): 1220–1223.

Chapter 5: Universe to Multiverse

1. A. Guth, *The Inflationary Universe: The Quest for a New Theory of Cosmic Origins* (New York: Basic Books, 1998), 15.

2. More precisely, Vilenkin found there was an energy barrier between these two regimes, so some range of radii for the universe was classically forbidden between the two. That meant that if only classical physics was at play, it would be very difficult to explain how a tiny universe could ever grow into a very large one because to do so it would need to—at some point—have one of these banned intermediate radii.

3. Technically, quantum laws should apply to larger objects too, but with less effect. That is, quantum tunneling is in theory possible for a large universe as well, but it is just extraordinarily unlikely to ever happen.

4. A. Vilenkin, "Creation of Universes from Nothing," *Physics Letters B* 117 (1982): 25.

5. It should also be mentioned that Stephen Hawking and American physicist Jim Hartle have proposed their own rival model of universe creation, and this too predicts the Bunch-Davies vacuum state.

6. If you're keeping track of how different inflationary models make different predictions for how density perturbations arose in our early universe, then it is worth noting here that in Vilenkin's model, it is the small variations between the inflationary paths of neighboring regions that causes these density perturbations.

7. A. Vilenkin, "The Birth of Inflationary Universes," *Physical Review D* 27 (1983): 2848.

8. J. Garriga and A. Vilenkin, "Many Worlds in One," *Physical Review D* 64 (2001): 043511.

Chapter 6: The Accidental Universe Makers

1. Our baby universe would also have to stop inflating at some point, when it reaches adulthood—just as our cosmos has—if it is to eventually generate

stars, galaxies, planets, and people. As we saw in Chapters 3, 4, and 5, Guth's, Linde's, and Vilenkin's models all hypothesized different ways that inflation might have ended in our universe, with some proposals working better than others. Even though there wasn't consensus on which of these models was most plausible, or whether further developments were needed, Guth and company assumed that since we do know that inflation must have ended somehow in our own universe, then if inflation could be kick-started in a baby universe in the lab—whichever mechanism was right—inflation would end in our lab-made cosmos too.

2. That telephone interview marked the only occasion when I have ever hung up on an interviewee. Vilenkin's tone was so serious that he sent me into a fit of hysterical laughter, and I was actually unable to speak. I put the phone down, composed myself, and sent him an apologetic email explaining that I'd had to end the conversation because I was laughing so hard that I actually couldn't breathe. "That's okay," he wrote back. "After I had said it, I realized it was kind of funny, and started laughing too."

3. Fans of Douglas Adams will recognize this as an homage to *The Hitchhiker's Guide to the Galaxy*.

4. Here I have outlined the traditional view of what happens to matter and light when it crosses the event horizon and is swallowed by a black hole: matter is eventually squashed into the singularity at its core. In recent years there has been some debate over whether matter would actually reach the core. A possible alternative is that matter is burned as it crosses the event horizon (see A. Almheiri, D. Marolf, J. Polchinski, and J. Sully, "Black Holes: Complementarity or Firewalls?" *Journal of High Energy Physics*, February 2013, 62). This question has yet to be resolved, but either way, black holes appear to be bad news for anything that is unfortunate enough to fall into them.

5. Some physicists have posited that all black holes, even cosmic ones, may house wormholes to other universes—and this idea will be discussed more fully in Chapter 9. But on the basis of this calculation, Guendelman, Guth, and Blau did not make that wider claim about all black holes, only the one they were examining.

6. That number, 25 grams or about 1 ounce, seems enticingly small. But Guth had also calculated that the density of that tiny patch of space would be 10^{80} grams per cubic centimeter. (For comparison, one of the densest cosmic objects spotted in the skies, excluding black holes, is the neutron star, which has a density of around 10^{14} grams per cubic centimeter.)

7. S. K. Blau, E. I. Guendelman, and A. Guth, "Dynamics of False Vacuum Bubbles," *Physical Review D* 35 (1987): 1747.

Chapter 7: A Baby Universe, a String Multiverse, or God?

1. The strings of string theory are not to be confused with the loops of loop quantum gravity that we encountered in Chapter 2. Strings live in spacetime, like elementary particles, whereas loops are proposed to create spacetime

itself. These theories are not connected, though whether they are complementary or incompatible is not yet known. And, of course, both could turn out to be wrong.

2. P. Davies, *The Goldilocks Enigma: Why Is the Universe Just Right for Life?* (London: Penguin Books, 2007), 163. This book contains numerous other detailed examples of fine-tuning in the cosmos and is well worth reading.

3. I have said that we "appear" to observe dark matter and dark energy because some physicists have attempted to construct alternative models in which these observations are caused not by the presence of mysterious invisible stuff that exists out there but by the fact that gravity can change its strength on different scales. These are not mainstream views, but they are valid and respectable ideas that are being investigated.

4. The discovery of dark energy was also good news for Guth, Linde, and other proponents of inflation (and consequently also for baby universe builders). Recall from Chapter 3 that inflation predicts that the universe is "flat"—that is, it has a certain critical density. When the astronomers estimated how much dark energy must be out there to account for the measured acceleration, they found that it makes up the missing 68 percent of the universe needed to reach this critical density. Inflation's prediction had been confirmed. Taken alongside the increasingly sophisticated satellite measurements of the cosmic microwave background, which continued to confirm the temperature patterns predicted by inflation, the theory was looking to be in good shape.

5. Einstein had actually inserted a cosmological constant into his own equations of general relativity. Back in Chapter 1, we saw that the equations naturally predict that the universe should be expanding (or perhaps contracting), and that it is unlikely to be static. Einstein found the prediction of a universe in motion to be too unsettling. He threw an extra variable into his equations, the cosmological constant, that would provide a constant outward push throughout the cosmos to counter the tendency for his static universe to shrink back inward if it was ever given a slight extra tug from the gravity of the matter it contained. When Einstein found out that the universe was actually expanding, he consigned the cosmological constant to the dustbin—but it seems to have returned with a vengeance less than a century later, with the discovery of dark energy.

6. This is not so different from the effect of an inflaton field creating inflation, as set out in Chapter 3. But an important difference is that inflation only lasted for a few fractions of a second, whereas the effects of a cosmological constant, if it exists, would be permanent.

7. At first blush, it might seem that eternal inflation, as laid out by Alex Vilenkin and Andrei Linde (see Chapter 5), would serve this purpose by automatically creating new cosmoses in their ever-growing multiverse. But as Guth realized when Vilenkin first presented him with the idea, there was no solid

reason at this point—it was the early 1990s—to think that the additional universes spawned by eternal inflation would have physical parameters different from those of their parent. So eternal inflation had sent Guth off to sleep, and it didn't set Polchinski's world afire either.

8. W. Fischler, D. Morgan, and J. Polchinski, "Quantum Nucleation of False Vacuum Bubbles," *Physical Review D* 41, no. 8 (1990): 2638–2641.

9. E. Farhi, A. H. Guth, and J. Guven, "Is It Possible to Create a Universe in the Laboratory by Quantum Tunneling?" *Nuclear Physics B*, 339 (1990): 417–490.

10. This is what Bryce DeWitt tried to do when he hit the problem of time, as laid out in Chapter 4.

11. If you're keeping track of the notes in this chapter, you'll remember that earlier I said that physicists "appear" to have measured dark matter and dark energy, but that there is a minority view that these entities don't exist. Instead, some physicists suggest that we need to modify our equations of gravity, and that doing so will explain away both effects. It's worth noting here that certain modified gravity theories may also make it easier to produce a baby universe in the lab. So it's not essential that string theory be correct to make baby universe production viable.

12. R. Bousso and J. Polchinski, "Quantization of Four Form Fluxes and Dynamical Neutralization of the Cosmological Constant," *Journal of High Energy Physics*, July 2000, 006.

13. Christoph Schönborn, "Finding Design in Nature," *New York Times*, July 7, 2005.

Chapter 8: The Little Bang Theory

1. K. Sato, "First-Order Phase Transition of a Vacuum and the Expansion of the Universe," *Monthly Notices of the Royal Astronomical Society* 195, no. 3 (1981): 467–479; K. Sato, "Cosmological Baryon-Number Domain Structure and the First Order Phase Transition of a Vacuum," *Physics Letters B* 99 (1981): 66–70.

2. K. Sato, H. Kodama, M. Sasaki, and K. Maeda, "Multi-Production of Universes by First-Order Phase Transition of a Vacuum," *Physics Letters B* 108 (1982): 103–107.

3. M. B. Einhorn and K. Sato, "Monopole Production in the Very Early Universe in a First-Order Phase Transition," *Nuclear Physics B* 180 (1981): 385–404.

4. A. Linde and D. Linde, "Topological Defects as Seeds of Eternal Inflation," *Physical Review D* 50 (1994): 2456–2468.

5. E. I. Guendelman and A. Rabinowitz, "The Gravitational Field of a Hedgehog and the Evolution of Vacuum Bubbles," *Physical Review D* 44 (1991): 3152–3158; A. Vilenkin, "Topological Inflation," *Physical Review Letters* 72 (1994): 3137–3140.

6. This idea that the monopole would appear to remain a point when viewed from outside but be seen to be inflating when observed from within follows a line of argument similar to that given by Blau, Guendelman, and Guth, as described in Chapter 6, so I won't repeat it again here.

7. A. Linde, "Monopoles as Big as the Universe," *Physics Letters B* 327 (1994): 208–213.

8. N. Sakai, "Dynamics of Gravitating Magnetic Monopoles," *Physical Review D* 54 (1996): 1548.

9. N. Sakai, K. Nakao, and T. Harada, "Causal Structure of an Inflating Magnetic Monopole," *Physical Review D* 61 (2000): 127302.

10. N. Sakai, K.-I. Nakao, H. Ishihara, and M. Kobayashi, "Is It Possible to Create a Universe out of a Monopole in the Laboratory?" *Physical Review D* 74 (2006): 024026.

11. In particular, Sakai and Guendelman need to find matter that has a negative mass.

12. Castelnovo was awarded the EPS CMD Europhysics Prize for the prediction and experimental observation of magnetic monopoles in spin ice in 2012 and the IUPAP C10 Young Scientist Prize in 2013.

13. There isn't a way to zoom in with a microscope and "see" which way spins are pointing with your eyes, since there aren't really any arrows there. But physicists use a technique called neutron scattering—essentially firing neutrons at a material to see how they bounce back, to build up a picture of what they hit. This is particularly sensitive to relative orientations of magnetic spins and can reveal spin ordering—or in this case, a lack of it.

14. C. Castelnovo, R. Moessner, and S. L. Sondhi, "Magnetic Monopoles in Spin Ice," *Nature* 451 (2008): 42–45.

15. J. Morris et al., "Dirac Strings and Magnetic Monopoles in the Spin Ice $Dy_2Ti_2O_7$," *Science* 326, no. 5951 (Oct. 16, 2009): 411–414, doi:10.1126/science.1178868 (2009); T. Fennel et al., "Magnetic Coulomb Phase in the Spin Ice $Ho_2Ti_2O_7$," *ScienceExpress Report*, September 3, 2009, doi:10.1126/science.1177582.

Chapter 9: Postcards from the Edge

1. The Very Early Universe meeting, University of Cambridge, 2007: www.damtp.cam.ac.uk/research/gr/workshops/VEU/2007/programme.html.

2. S. Hawking, *A Brief History of Time* (New York: Bantam Dell, 1988).

3. D. N. Page and S. W. Hawking, "Gamma Rays from Primordial Black Holes," *Astrophysics Journal* 206 (1976): 1–7.

4. Matthew 8:28–34.

5. Luke 17:34, Matthew 24:40.

6. See in particular A. Almheiri, D. Marolf, J. Polchinski, and J. Sully, "Black Holes: Complementarity or Firewalls?" *Journal of High Energy Physics*, February 2013, 62, which has led many physicists to reexamine the fate of objects and information falling into a black hole.

7. J. Garriga, A. Vilenkin, and J. Zhang, "Black Holes and the Multiverse," *Journal of Cosmology and Astroparticle Physics*, 2016, doi:10.1088/1475-7516/2016/02/064.

8. The team also examined another mechanism in which black holes are formed inside spherical domain walls. Reaching back to concepts explored in Chapter 3, a domain wall is like a fracture in space that formed as the universe cooled, just like defects form as the crystal structure of ice cubes misalign, making them appear cloudy.

9. The team has investigated a number of different versions of this model, as mentioned in the previous note, and the value of the critical mass varies depending on the version they are looking at. It can reach up to millions of times the mass of the Sun.

10. Since Vilenkin's team first formulated their model, the LIGO collaboration announced the discovery of gravitational waves from the collision of two black holes, as discussed in Chapter 2. Data from such gravitational waves can help astronomers gauge how massive the colliding black holes must have been. Vilenkin contends that as more gravitational wave events (that is, more such cosmic smash-ups) occur, astronomers will be able to map out the distribution of black holes with different masses. These data could be used to test the model over the next few years.

11. S. Dimopoulos and G. Landsberg, "Black Holes at the LHC," *Physical Review Letters* 87 (2001): 161602.

12. K. Larsen and R. L. Mallett, "Radiation and False-Vacuum Bubble Dynamics," *Physical Review D* 44 (1991): 333.

13. N. Arkani-Hamed, S. Dimopoulos, and G. Dvali, "The Hierarchy Problem and New Dimensions at a Millimeter," *Physics Letters B* 429, nos. 3–4 (1998): 263–272.

14. S. B. Giddings and S. D. Thomas, "High-Energy Colliders as Black Hole Factories: The End of Short Distance Physics," *Physical Review D* 65, no. 5 (2002): 056010.

15. S. B. Giddings and M. M. Mangano, "Astrophysical Implications of Hypothetical Stable TeV-Scale Black Holes," *Physical Review D* 78 (2008): 035009.

16. There are discussions to build an International Linear Collider, at a location as yet undecided, and possibly another circular collider that would smash together muons—heavier cousins of the electron—at a new accelerator at Fermilab. These are also in the planning stages, but would have lower energies than the LHC (though they would offer other advantages in the kinds of measurements that could be made).

17. I should also emphasize, as Guendelman—who is not a particular fan of string theory—has stressed to me on many occasions, that the extradimensional paradigm is not the only exotic hypothesized framework for bringing the scale of baby universe fabrication within our reach. It is simply the best-known and the most widely investigated at particle accelerators. But even if string

theory proves to be wrong and/or extra dimensions do not exist, there are other routes to building a universe in the lab that Guendelman and his colleagues are exploring, some of which I touched on in Chapter 8.

18. D. N. Page, "A Theological Argument for an Everett Multiverse," December 21, 2012, arXiv:1212.5608v1.

Chapter 10: Living in the Matrix

1. S. Lloyd, "Computational Capacity of the Universe," *Physical Review Letters* 88 (2002): 237901.

2. "We may be able to fit humans into our simulation boxes within a century." Silas Beane, quoted in Z. Merali, "Do We Live in the Matrix?" *Discover*, December 2013.

3. N. Bostrom, "Are You Living in a Computer Simulation?" *Philosophical Quarterly* 55, no. 211 (2003): 243–255.

4. J. D. Barrow, "Living in a Simulated Universe," in *Universe or Multiverse?*, ed. B. Carr, 481–486 (Cambridge: Cambridge University Press, 2007).

5. J. K. Webb et al., "Search for Time Variation of the Fine Structure Constant," *Physical Review Letters* 82, no. 5 (1999): 884–887.

6. S. R. Beane, Z. Davoudi, and M. J. Savage, "Constraints on the Universe as a Numerical Simulation," *European Physical Journal A* 50 (2014): 148.

7. S. Lloyd, "A Turing Test for Free Will," *Philosophical Transactions of the Royal Society A* 28 (2012): 3597–3610.

8. A. Sandberg, "Transhumanism and the Meaning of Life," in *Religion and Transhumanism: The Unknown Future of Human Enhancement*, ed. C. Mercer and T. Trothen, 3–22 (Santa Barbara, CA: Praeger, 2014).

Index

Zeeya Merali is an award-winning freelance science journalist, author, documentary maker, and editor for the Foundational Questions Institute, FQXi. She has written for *Scientific American*, *Nature*, *New Scientist*, *Discover*, *Science*, and the *Wall Street Journal*, has published two textbooks in collaboration with National Geographic, and has produced programs for BBC radio and History UK. She holds first-class undergraduate and master's degrees in natural sciences from the University of Cambridge and a PhD in theoretical cosmology from Brown University. She lives in London. Visit her website at thelittlebangtheory.com.